IN SEARCH OF ANCIENT ASTRONOMIES

IN SEARCH OF ANCIENT ASTRONOMIES

Edited by E. C. Krupp

Doubleday & Company, Inc., Garden City, New York

ACKNOWLEDGMENTS

The editor, contributors, and publisher express their appreciation to the following for permission to include:

Numerous original illustrations, as well as those redrawn from other sources, as noted in the text, courtesy of the Griffith Observatory, Los Angeles, California.

Numerous Griffith Observatory redrawings of illustrations from two books by Alexander Thom, published by Oxford University Press: *Megalithic Sites in Britain* (1967) and *Megalithic Lunar Observatories* (1971). Courtesy of the author and Oxford University Press.

Page 60, Griffith Observatory redrawings of illustrations from "Archaeological Tests on Supposed Prehistoric Astronomical Sites in Scotland," by Euan W. MacKie, in *Philosophical Transactions of the Royal Society of London*, Vol. 276 (1974). Reprinted by permission.

Numerous Griffith Observatory redrawings of illustrations by Alexander Thom from the *Journal for the History of Astronomy*, Vols. 2,3,5,6. Courtesy *Journal for the History of Astronomy*.

Page 78, a Griffith Observatory redrawing of an original illustration from "The Astronomical Significance of the Crucuno Stone Rectangle," *Current Anthropology*, Vol. 14, by A. S. Thom, by permission of The University of Chicago Press. © 1973 by the Wenner-Gren Foundation for Anthropological Research. All rights reserved.

Pages 94, 99, and 103, Griffith Observatory redrawings of original illustrations from *Stonehenge Decoded*, by Gerald S. Hawkins and John B. White. Copyright © 1965 by Gerald S. Hawkins and John B. White. Reprinted by permission of Doubleday & Company, Inc.

Page 96, a Griffith Observatory redrawing of an illustration from *Wessex Before the Celts*, by J. F. S. Stone. Courtesy of Thames and Hudson Ltd., London.

Pages 108–9, Griffith Observatory redrawings of illustrations from *From Stonehenge to Modern Cosmology*, by Fred Hoyle. W. H. Freeman and Company, San Francisco. Copyright © 1972. Used by permission.

for Florence and Edwin Krupp,
my mother and father,
who appreciate both
a good mystery
and common sense

Contents

ACKNOWLEDGMENTS: E. C. KRUPP

I am very appreciative of the contributions by Alexander Thom, A. S. Thom, John Eddy, and Anthony Aveni to this book. All are recognized experts in their fields and obviously did much to make this book worthwhile. All of them are gentlemen, with very human-oriented approaches to the cosmos, and I value their friendship highly. I am especially grateful to Professor Alexander Thom for his entertaining conversations and illuminating correspondence.

R. J. C. Atkinson, Horst Hartung, the late C. A. Newham, and Margot Wood were all very generous with their time and sage advice.

This book also owes much to Griffith Observatory, Department of Recreation and Parks, City of Los Angeles, and I am grateful for permission to reproduce some of the many fine illustrations in the observatory's file, many of which were prepared by John Lubs and Hugh Johnson.

I must also thank Jane Jordan Browne, who courteously dogged this book to its end, and Alex Liepa, who provided, light-handedly, many suggestions for improvement of the manuscript, and Georgiana Remer, whose copyediting had a very salutary effect on the book. I thank Margaret Rector, whose annual Christmas message one year did much to get the book going, and Robert Rector, for his continuous support. For the same reason I must express my appreciation to my parents, Edwin and Florence Krupp, and to my wife, Robin, and son, Ethan, who tolerated all of the usual disappointments precipitated by a book in the works.

Finally, special thanks and recognition are due to UCLA Extension's Department of Biological and Physical Sciences and, in particular, to Dr. Robert Barrett and Carole Whittaker, who stimulated the development of the course without which this book could not have evolved.

ACKNOWLEDGMENTS: JOHN A. EDDY

I am indebted to the following colleagues for assistance and helpful comments: John Brandt, Von Del Chamberlain, Thaddeus Cowan, Richard Forbis, Ernest Hildner, Alice Kehoe, Thomas Kehoe, Edwin Krupp, Jonathan Reyman, Payson Sheets, Waldo Wedel, and Warren Wittry. Some of my work on medicine wheels cited here was supported by the National Geographic Society.

ACKNOWLEDGMENTS: ANTHONY F. AVENI

The author gratefully acknowledges Dr. Sharon L. Gibbs for valuable comments made on the manuscript and Arqto. Horst Hartung for providing several illustrations.

Introduction

At Stonehenge in England and Carnac in France, in Egypt and Yucatán, across the whole face of the earth are found mysterious ruins of ancient monuments, monuments with astronomical significance. These relics of other times are as accessible as the American Midwest and as remote as the jungles of Guatemala. Some of them were built according to celestial alignments; others were actually precision astronomical observatories. All are wordless but emphatic evidence of our ancestors' energetic pursuit of the sky and stars. They mark the same kind of commitment that transported us to the moon and our spacecraft to the surface of Mars. Careful observation of the celestial rhythms was compellingly important to early peoples, and their expertise, in some respects, was not equaled in Europe until three thousand years later.

We are just awakening to the success achieved in astronomy by many ancient and prehistoric peoples. We are just beginning to appreciate their perseverance, for it must have taken much more than a single generation to establish some of the subtler cycles in the sky. This re-evaluation is also reminding us that these people were bright and curious and are deserving of our respect.

Now, in an era of wild, unsubstantiated claims about ancient astronauts by Erich von Däniken and his fellow travelers, their pseudoscientific misconceptions of these earlier peoples are in part dispelled by the reliable, scientific findings of archaeoastronomy.

We do not always trouble ourselves to notice the same things in nature that were of life-and-death interest to our ancestors. What we make of what we see is also changed. In earlier times the sky was a place. Now it is just a direction. But back when the sky was not just an illusion in blue but a sphere of activity, people noticed the more obvious things and, like ourselves, fashioned an understanding of the cosmos from them. Sometimes this view was expressed in their monumental architecture, and because we are now beginning to sense this, some of the most tantalizing archaeological mysteries are being explained. A new science has evolved in the process.

Archaeoastronomy is the study of the astronomies of ancient and prehistoric times. We say "astronomies" because, unlike today, observational approaches and astronomical interests varied from place to place and from time to time. The Megalith Builders of Western Europe, at high northerly latitudes, made precise sightings on the horizon. In ancient Mexico, far to the south, passages through the zenith were also observed. Both the similarities and the differences in the ways science was conducted in its earliest days shed light on how we have come to be the way we are.

Archaeoastronomy is a truly interdisciplinary science. At its best it combines the skills and background of modern archaeology with the numerical certitude of practical astronomy. Its practitioners number a couple of dozen at the most, and each of these individuals has helped develop this new science from the ground floor. There are no university departments of archaeoastronomy. As a result, most of those who contribute to it have arrived by different routes. Some are astronomers. They have an insight into the phenomena that would interest early astronomers, and they apply their mathematical techniques to deduce what alignments were constructed. They determine what calculations were performed. Archaeologists are able to evaluate the details of an individual site. They date the site and reconstruct its history. They translate old inscriptions. They obtain independent evidence which is needed to confirm the astronomical hypothesis. Ethnographers search for clues in ancient customs and in those that have persisted to the present day. Still other scholars, experts in applied mathematics or engineering or architecture or whatever else is needed, apply their special talents to the archaeoastronomical problem.

The newness of archaeoastronomy is part of its appeal. It offers ample opportunity for serious researchers to make important contributions. New discoveries appear regularly in the scientific journals. All of this activity generates considerable excitement and keeps the communication between archaeoastronomers lively. It is a controversial science; some of its results call long-standing beliefs into question. But it contains within itself the means to separate fact from fancy, conclusion from hypothesis, and deserves to be taken seriously.

This book is the first attempt to present systematically to the general reader the main results of archaeoastronomy to date. There have been some popular books on some of the sites mentioned in the following pages, but up until now it has not been possible to obtain a balanced perspective of the whole field. Most of the work was scattered through a variety of scholarly publications, and it was difficult, even for an archaeoastronomer, to see the entire picture.

In order to assemble a comprehensive account of archaeoastronomy today, several of its foremost experts have contributed chapters in their respective areas of interest. This book grew out of a special series of lectures, "In Search of Ancient Astronomies," first presented in the winter of 1975 at the University of California Los Angeles and the University of California San Diego and developed through UCLA Extension's Department of Biological and Physical Sciences. The contributors here participated in that series.

Many who are interested in archaeoastronomy, both general readers and professional scholars, may not be familiar with spherical astronomy. In the first chapter, therefore, a non-mathematical description of the behavior of the sun, moon, and stars is presented to introduce practical astronomy to the reader and to illustrate, with examples from various cultures, how observations of fundamental events may have been made. The daily and yearly patterns of the sun, the monthly cycle of the moon, and even subtler motions of the moon and stars are all described in detail.

Having mastered the basics of astronomy, we are introduced to some of the earliest sites of all, the Megalithic ruins in Britain and France, by the world's leading archaeoastronomer, Professor Alexander Thom, now retired from Oxford University, and his son, Dr. A. S. Thom. Professor Thom has spent the last five decades carefully measuring and interpreting the old stone rings and standing stones of Britain and France. His work demonstrates an unexpected interest in geometry that seems to have gone hand-in-hand with Neolithic astronomy. Thom's work has also forced us to consider the origins of such practical things as the unit of measurement. Evidence from Carnac, Argyll and the Orkney Islands of Scotland, from over three hundred personally surveyed sites, is summarized here. Chapter 2 demonstrates why archaeologists are now reassessing the nature of science and society in prehistoric Britain in the light of the Thoms' research.

Of all the Neolithic and Bronze Age monuments, Stonehenge is the best known and most evocative. Stonehenge astronomy, as popularized by Professor Gerald Hawkins in 1963 and by Sir Norman Lockyer before him, has been a great stimulus to the growth of archaeoastronomy. Until now, however, a complete account of the astronomical investigations of Stonehenge has not been available, but in the third chapter the contributions of well-known investigators and others, also important but not as well known, are explained. This account concludes with the new Stonehenge astronomy of Professor Thom and demonstrates that the moon may have been of even greater interest to the Stonehengers than the sun.

It is a long leap from Stonehenge to North America, but Dr. John A. Eddy proves in Chapter 4 that the archaeoastronomy of the American Indian shares some of the difficulties we have with Stonehenge astronomy. Comparisons have been drawn between the social organization of Bronze Age Britain and tribes in the Mississippi Valley. Although the North American Indians left no written records of their activities, recent research indicates that they were interested in some of the celestial phenomena that were observed much earlier by the Megalith Builders. The Indians observed these things and recorded them in other ways.

Some of their records are unique. The Anasazi in the southwestern United States seem to have recorded pictographically the supernova explosion of A.D. 1054. John Eddy relates these results and describes in detail what kinds of astronomical structures still exist in the ruins of the Southwest, among the mound builders' urban centers, and on the Great Plains. Eddy's own work on the Big Horn Medicine Wheel in Wyoming is included, and he brings us up to date on his most recent research on other medicine wheels in the American West.

South of the border in Mexico, Guatemala, and Honduras, the peoples of Mesoamerica developed an intricate and amazingly accurate astronomical calendar. Their few remaining writings also allude to careful astronomical observations and contain tables of astronomical data. One of the greatest mysteries of Mesoamerica is this astronomical expertise. We know it was there, but how was it achieved? Dr. Anthony F. Aveni has personally surveyed many Mesoamerican sites and is in a unique position to narrate the story of archaeoastronomy in Mexico and Central America, as he does in Chapter 5. We discover that the entire street plan of Teotihuacán, the "Home of the Gods," was laid out according to the stars. Curiously shaped buildings in Oaxaca and Yucatán contain astronomical alignments and actually may have been functioning astronomical observatories. Dr. Aveni's own participation in this research brings this narrative to life.

The highly organized society of ancient Egypt was unique in the Middle East in developing a civil solar calendar. This very practical invention eventually evolved into the tropical calendar we use today. Egyptian astronomical techniques also led directly to the familiar convention of the twenty-four-hour day. Many more written records from the ancient civilization of the Nile remain than can be found in Mesoamerica, yet despite this, we are still trying to discover the full range of Egyptian astronomy. We are still trying to learn how they made their observations. A variety of unambiguously astronomical relics have survived: coffin-lid star clocks, merkhets and bays (plumb lines and slotted sticks), tomb paintings, and temple reliefs. The controversial temple alignments of Lockyer and in Hawkins' more recent work and the voluminous literature on the pyramids, some accurate and some completely contrived, add to the picture. In Chapter 6 what we know of Egyptian archaeoastronomy is described.

This book's emphasis is primarily upon the archaeology and astronomical interpretation of monumental architecture. The pretelescopic astronomies of Mesopotamia, China, Hindu India, and even ancient Greece, however, are also interesting chapters in the story of ancient astronomy, but what we know of these other approaches largely is based on written records. Much of what we see of their science represents an

already well-established tradition, and discussion of them here would di-
lute the main theme of this book. Doubtless exciting work will continue
to reveal new insights into the origins of these and other astronomies
found, for example, in southeast Asia and Polynesia, and in time a companion to the present book may be possible.

Archaeoastronomy will by no means solve all the mysteries of the
past, but it will help provide a fuller picture of the lives of those who
have lived before us. In the meantime, however, unsubstantiated chal-
lenges to modern science—Immanuel Velikovsky's catastrophes, Alfred
Watkins' ley lines, Katherine Maltwood's terrestrial zodiac, Erich von
Däniken's ancient astronauts, to name a few—have secured a wide
readership and generated interest and support. Astronomy and archaeology both have romantic appeal, and for this reason, perhaps, both
have been exploited in the name of many pseudoscientific fantasies.
Several of these topics are critically examined in the last chapter of this
book, to provide, by contrast, an understanding of the difference between true scientific inquiry and fanciful speculation.

If we are going to find the roots of astronomy in the enigmatic monuments and incomplete records left by ancient and prehistoric peoples,
and that is the itinerary of this book, then we are going to have to mar-
shal all our resources of common sense and logical thought and seek
out the things our ancestors were watching.

IN SEARCH OF ANCIENT ASTRONOMIES

CHAPTER ONE

A Sky for All Seasons

E. C. KRUPP

For the peoples of antiquity the sky was always overhead. What happened there repeated itself, and these repetitions made it possible to structure time and the world, as they do for us today.

Some of the cycles are simple. Others are complex and difficult to master, and today, when we are removed by city lights and air pollution from a clear view of the sky and are distracted by our daily affairs, most of us are not even aware that the cycles exist.

The sun persists in its daily risings and settings, but through the year the sun's first and last appearances vary in time and place. In consequence, we have seasons. The moon, too, alters its face, waxing and waning through the month. The moon has far subtler motions also, however, and had we time and inclination, we could eventually discover them.

Although the sun and the moon go through their changes, the stars at least seem ever constant. But the stars, too, change, from season to season, and through the long cycle of precession, even their seasons change.

In this chapter the daily and annual motion of the sun is explained and related to the observations the early astronomers made. The moon's monthly pattern of phases, its 18.61-year cycle of regression, and its tiny wiggle of a perturbation are also described and explained, as are eclipses, which depend on these lunar variations. The progress of the stars, too, is explained, and particular attention is given to the phenomenon of heliacal rising.

We have always looked to the heavens for orientation and perspective. It satisfies a need. It may be what our brains require to perceive the world at all. Anything might do the job, but the heavens do it well. They repeat themselves over and over, and no one can tamper with them.

THE SIMPLE SUN

The sun, the brightest object in the sky, exhibits a daily motion. It rises roughly in the east and sets in the west, as do all celestial objects. This apparent motion is due, of course, to the rotation of the earth on its axis, but from the purely observational point of view of the ancient astronomers it also hardly matters. The important thing is the phenomenon itself, and we are looking for evidence of its having been observed.

The cardinal directions—north, south, east, and west—are devices with which we orient the world. They are based upon two fundamental places of reference, the horizon and the zenith. The horizon is where the earth meets sky, and it surrounds the observer. Ideally it is the circular boundary of the sky at ground level for an observer located at the circle's center. Naturally, foreground objects often obscure the true horizon, but it is still possible to imagine it without difficulty. The zenith is the point in the sky directly overhead, directly above the observer. These two concepts, a circle and a point, constitute a very personal reference frame. In this very democratic cosmos everyone has his or her own horizon and zenith. If one moves ever so slightly, the bounds of one's horizon and the direction of one's zenith change.

The cardinal directions are all located on the horizon. It is as if someone had painted or lighted signs out on the horizon to indicate those directions. They would have no meaning were the earth not in rotation. The earth rotates from the direction we call "west" to the direction we call "east." Because we are rotating with the earth and because everything else on the earth is rotating with us, we have no sensation of motion. The sun, moon, and stars all move around, to be sure, but their apparent motion is sufficiently slow to deceive us.

Once east and west are established, say, by the risings and settings of celestial objects, it is a simple matter to go halfway between them and call one direction north and its opposite south. North has a special meaning, though, because we see the earth's rotation reflected in the sky. In the northern hemisphere it looks as though the whole sky is spinning around a single point. The point is called the "north celestial pole," and it is just a direction in space, now roughly toward the star Polaris, toward which the earth's rotational axis points.

The earth is roughly spherical in shape, and this means that one's movement to the north or south will change the zenith. The observer's view of the heavens is therefore altered by his movement along the earth's surface. In particular, as an observer moves north, the north celestial pole moves higher in the sky. As an observer moves south, it moves closer to the horizon. This makes sense. If one travels far enough north to stand on the earth's north pole, the sky's north pole will be directly overhead. All of the visible stars will appear to circle around the celestial north pole, as the earth rotates, and all visible stars will remain above the horizon.

To all but the most precise instruments of measurement, the rate of the earth's rotation is extremely uniform and constant. This uniformity produces an apparent sequence of repetitive events in the sky—sunrises, for instance—and permits the measurement of time through celestial observations.

We perceive the sky over our heads as the inside surface of an inverted bowl, or hemisphere, which arcs down in every direction to the horizon. In addition to the horizon, the local celestial meridian is another convenient reference for description of celestial phenomena. Any northern hemisphere observer has a local celestial meridian which is the circular arc that passes from the north point on the horizon through the north celestial pole, through the zenith, and on to the south point on the horizon. All celestial objects appear to cross, or transit, the celestial meridian as the earth turns. As they cross the local celestial meridian, they also reach their highest angles, or altitudes, in the sky. When the sun transits the local celestial meridian, the local apparent solar time is said to be noon, and this usually occurs close to the time our clocks read twelve noon. If we imagine that our sky looks like a hemisphere suspended over our heads, we can also imagine a second, invisible hemisphere that is below the ground and that connects with the visible sky, all around the horizon. This second hemisphere is, at any moment, the unseen part of the sky, and it is unseen, naturally, because the earth is in the way. The earth's rotation eventually brings at least part of the unseen sky above the horizon and into view. The part of the local celestial meridian that extends around the unseen hemisphere of the sky is called the "lower branch" of the meridian, and it passes through the south celestial pole. This lower branch is the local celestial meridian for observers in the opposite hemisphere with this meridian, and when the sun transits the lower branch, for them it is noon. For those in the north, who use its upper branch, it is midnight.

After sunrise but before the sun transits the upper branch of a time zone's standard celestial meridian, the clock time is specified as A.M., or ante-meridiem ("before midday"=before meridian transit). After the sun transits the upper branch of the meridian, the clock time is said to be P.M., or post-meridiem ("after midday"=after meridian transit). These terms are all part of our everyday experience and are in use, of course, simply because the earth rotates on its axis. This daily motion is responsible for the cycle of night and day and for the diurnal rising and setting of each celestial object.

A second apparent motion of the sun provides a second unit of time. The annual solar movement defines the year just as the diurnal motion

The sky can be imagined as a huge, hollow sphere at the center of which is the earth, but it is impossible to draw the earth and this imaginary celestial sphere to the same scale. The earth is very small compared to the distances of stars. If we imagine the earth as just a point in this picture, we can visualize how an observer at the top of the earth, at the north pole, sees his zenith straight overhead on the celestial sphere and also how he sees half of the entire sky. The surface of the earth, upon which he is standing, blocks his view of the bottom half of the sky. The horizontal circle on the celestial sphere therefore corresponds to the limits of the observer's horizon. (Griffith Observatory, John Lubs)

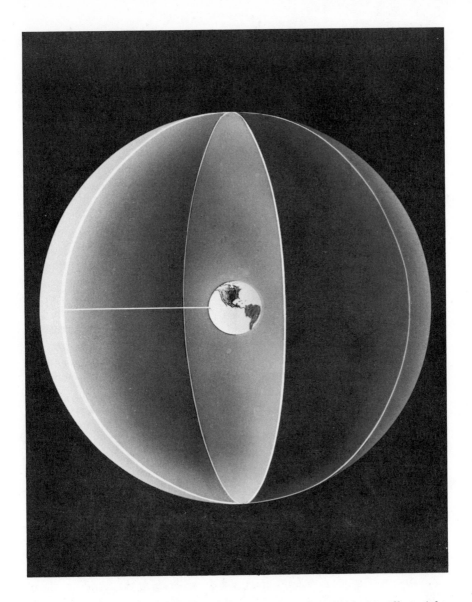

For an observer located on the earth's equator, the zenith is still straight overhead, but from our point of view outside the celestial sphere this direction is to the left. The celestial sphere is again divided into two halves by the observer's horizon, and in this case only the left portion of the celestial sphere is visible to our astronomer at this time. (Griffith Observatory, John Lubs)

defines the day. If it were possible to see the sun and the stars in the sky at the same time, the sun would be observed to move slowly across the background stars along a well-defined path. The movement of the sun can be detected indirectly through observation of the slowly changing composition of the nighttime sky. Some patterns of stars, or constellations, are visible in the evening summer sky, but as the year progresses, the sun moves into that region of stars. These same stars then rise and set with the sun and are invisible because the scattered sunlight of the daytime sky is so bright. The constellations are obvious keys to the seasonal cycle, too. Orion, for instance, is prominent through winter nights, but Scorpius dominates the summer nights.

The annual celestial circuit of the sun is due to the motion of the earth in its orbit around the sun. On earth we do not sense this motion. Instead, it appears as if the sun moves in relation to the background of stars. Even though we cannot see the daytime stars, they can be charted, along with the sun's position among them, by timing the sunset and watching for the first stars to appear in the west on successive evenings at the same specified time after sunset. This is a laborious procedure, and some small errors will creep in because the earth's orbit is not exactly circular, but gradually the technique will permit a good understanding of the annual motion of the sun. In a planetarium, by contrast, we can observe the sun and stars simultaneously to see easily that through the course of a year the sun moves completely around a circle in the sky. The circle is called the "ecliptic." The constellations of stars that fall along it are known collectively as the "zodiac," a word which means "ring of animals."

Alert observers would soon notice a pattern in the sun's motion that is related to the earth's orientation in space. During the course of the year the duration of hours of daylight varies from a minimum in winter to a maximum in summer. Autumn and spring occur at times when the hours of daylight and night are roughly equal. Even though most of us are not aware of the day-to-day changes in appearance and position of the sun, moon, planets, and stars, we do still sense the yearly cycle of the changing number of daylight hours. We also sense a special quality in the summer daylight, in which the hot sun is high and bright and the shadows are short. In winter the sun is lower in the sky throughout the day. The air is crisper. The windows on the northerly sides of the house are not so well lit. The sunlight seems diffused, and the shadows are long. To mark these changes in the seasons we might use the cycling height of the sun at noon: maximum in summer, minimum in winter. We would also notice that the sun's points of rising and setting vary through the year. On the longest day of the year, which we call the first day of summer (in Britain, midsummer), the sun rises at a point farthest to the north and sets likewise. On the first day of winter, the time of shortest daylight, the sun rises and sets at points to the farthest south. On the first days of spring and fall, the sun rises and sets due east and west respectively.

The seasonal changes of the sun, indeed, the seasons themselves, result from the earth's orientation in space. The earth's axis of rotation is tipped with respect to the direction of its orbital motion around the sun. We do not know what specific events or processes led to the present orientation of the earth's axis at an angle of 23½ degrees, but that element of primordial history created circumstances to which living creatures, including humans with their seasonal rituals, have been responding ever since.

Vegetation growth cycles are related to the seasons. Agricultural techniques are related, in turn, to the growing cycles. Farming provides food, necessary for life, and recognition of these relationships is expressed by many cultures in rituals. Holidays and their feasts and sacrifices are bound to the seasonal round. They define the calendar's circuit, and the calendar is perfected to establish them.

Some holidays occur at astronomically significant times. Astronomical phenomena are both indicators of the orientation and motion of the earth in space and, in turn, the indicators of the changing seasons. To ancient peoples it might have seemed that the astronomical indicators were simultaneously the cause and symbol of the world's great forces. Observation of the astronomical indicators would allow changes to be charted and anticipated.

One class of astronomical monuments includes alignments that indicate at least one particular direction on the horizon. Significantly, some object, usually the sun or the moon, appears there at a special time. It is really the occurrence of the celestial event, of course, that makes that time of singular interest. This may be the moment of sunrise or sunset on the longest or shortest day of the year. These days, the first days of summer and winter (or "midsummer" and "midwinter," in Britain) astronomically go by the names summer solstice and winter solstice. These same names are applied to the time of their occurrence as well. The summer and winter solstices occur roughly near the 21st of June and December, respectively, and mark extreme positions of the sun. The first days of spring and fall occur midway between the solstices, roughly on the 21st of March and September.

If we only were a little more immersed in the natural environment, we could not miss the solstices, those times when the sun seems to stop in its tracks and double back to where it appeared on the mornings before. Each day approaching the solstice the sun rises a little closer to the extreme position, to the north in summer and to the south in winter.

The amount of movement looks less each day until the sunrise is stopped in its motion along the horizon. For a few days the sun appears to linger at the same extreme at dawn. From this behavior derives the term "solstice," which means "sun still." The date of this occurrence could be marked by putting up a monument to point out the direction of the sun on the horizon on that special day at rising or setting.

Likewise, we would know the equinoxes well. On the equinoxes, the vernal in spring, the autumnal in fall, which mark the halfway points between summer and winter, the duration of daylight is the same as that of night. The meaning of the word "equinox" is, sensibly, "equal night."

The changing position of the sun throughout the year can be charted on the celestial sphere, and it is observed to oscillate above and below the celestial equator. Just as the north celestial pole is an extension of the earth's north geographic pole to the celestial sphere, the celestial equator is the extension of the earth's equator. At the winter solstice the sun is at its greatest angle below the equator. The northern hemisphere receives the sunlight obliquely, the weather is cold, and the sun rises and sets to the south. At the equinoxes the sun crosses the equator, and the weather is in transition. In summer, the sun is as far north of the celestial equator as it gets. The northern hemisphere receives direct sunlight, the weather is hot, and the sun rises and sets to the north.

In the sky these effects are manifested by the angle between the ecliptic, the path to which the sun is constrained, as reflected by the earth's orbital motion, and the celestial equator, whose movement mirrors the earth's rotation. These two circles are set at angle to each other because, again, the earth's axis is tilted, and this tilt of 23½ degrees is called the "obliquity of the ecliptic."

Evidence of solar alignments that are useful observationally comes from locations as widely separated as Wyoming and Scotland. We shall see that the Cahokia Mounds, laid out by Amerindian mound builders near present-day East St. Louis, Illinois, include a device that may have permitted a determination of the date of the solstice. Maya date glyphs imply that this Mesoamerican Indian culture had established the length of the tropical (or solar) year, the time from summer solstice to summer solstice (365.24220 days) with considerable accuracy.

The ancient peoples who committed themselves to ambitious programs of astronomical observation are separated widely by geography and time, but they usually shared at least one attribute: they were settled on land on which they built entire complexes of permanent public structures. Their efforts in earth and stone indicate a high level of social organization, with all the central authority and division of labor that may imply.

It is no coincidence that wealthy, agrarian societies like those in Egypt and the Americas spared no effort to make observations of the sky. Nor is it surprising that observatories were included among their major structures. Without the practical benefits that astronomy supplied, it might not have been possible to have civilization at all.

The rhythm of life is the rhythm of a culture. This rhythm is keyed to the seasonal cycle, as is the yearly agricultural cycle. A calendar is the expression of our sensitivity to these cycles. The calendar is a practical device, and its immediate application to agriculture is obvious. Yet the real power of the calendar goes beyond this. It is the device that permits complex organization of a culture, the device that rules the exchange of goods and services.

If there are farming surpluses, time becomes available to people for other activities which in turn enhance the society's economy. If full-time farming is not required of all society, division of labor is stimulated. More goods and services become available, and the society grows more complex. The increasing complexity demands a precise calendar. A small farming effort wouldn't require this. Even today, backyard gardeners can rely on natural woodlore and folk traditions to cue them to the appropriate time for planting and harvesting.

Our economy is stitched to the sky. As the scale of our agricultural enterprise increases, it is important that a device as common to us as the calendar be available to maintain smooth operations and the social order. We take it for granted that we can make appointments, schedule vacations, and remember birthdays with such ease. We are removed by technology from an everyday awareness of how and why our calendar works. Most of us do not heed the height of the sun at noon or the azimuth of its rising on any particular day. Nor are we particularly aware of the moon's phase unless we should chance to see it. In the past we were closer to the source of our food than to the all-night market, and we were more aware of celestial phenomena.

This awareness changed as society grew more complex. Responsibility for the calendar fell to a more-and-more specialized class of astronomer-priests, and ritual overtook in importance the celestial event prompting it.

There are still relics of ancient calendrical traditions in certain holiday customs. The ritualization of obvious astronomical events into holidays and ceremonies oriented the community to its common needs and purposes. Even today it is no coincidence that Christmas falls so near the winter solstice. There is still a bit of the vegetation myth in Yule celebrations.

A reliable calendar can be obtained by observing the rising (or setting) points of the sun. The solstices mark the two extremes between

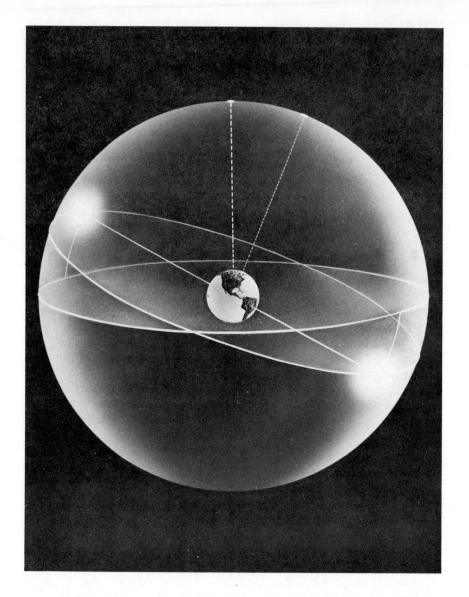

Two other circles centered on the earth and traced out on the celestial sphere are important. One of these is the celestial equator and the other is the ecliptic. In this case, the celestial equator is drawn horizontally, and the north celestial pole is straight overhead, above the north terrestrial pole. Summer solstice occurs when the sun is at its highest point above the celestial equator. The winter solstice takes place when the sun is at its lowest point below. The equinoxes occur at the two intersections of the celestial equator and the ecliptic. (Griffith Observatory, John Lubs)

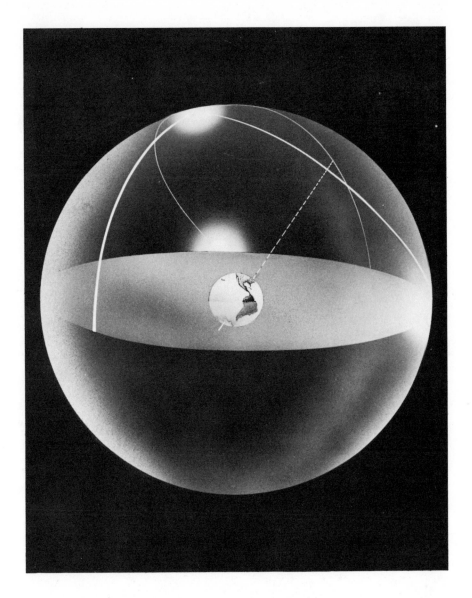

At the summer solstice, in the northern hemisphere, the sun rises as far to the northeast as it ever will. During the day, at noon, it transits as high above the horizon as it is ever seen. Finally, the sun sets as far to the northwest as it ever does, on this, the longest, day of the year. (Griffith Observatory, John Lubs)

When either equinox arrives, the sun rises due east, arcs across the sky at a height midway between those at the solstices, and sets due west. (Griffith Observatory, John Lubs)

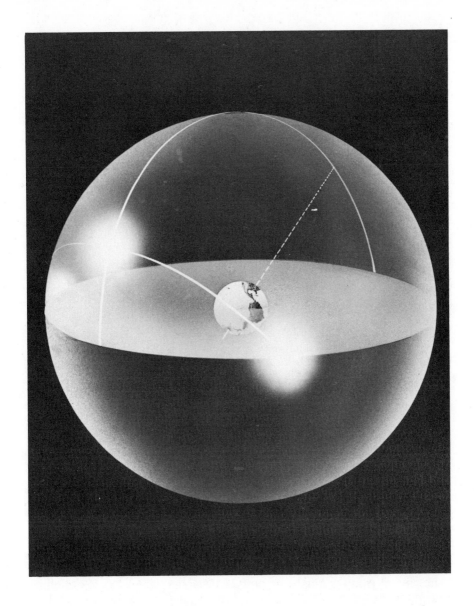

About six months after the summer solstice, the winter solstice takes place. Now the sun rises, as seen from the northern latitudes, in the southeast. It transits at its lowest altitude of the entire year, and it sets at its extreme position to the southwest. (Griffith Observatory, John Lubs)

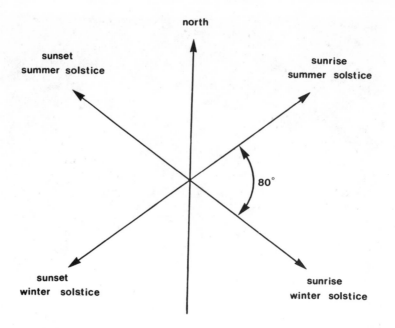

Extreme Azimuths Of The Sun At Stonehenge

At the latitude of Stonehenge, in Wiltshire, England, 51 degrees north, six months shift the rising and setting points of the sun at the summer solstice 80 degrees north of what they had been at the winter solstice. This shift is obvious and easily observed throughout the cycle of the year. (Griffith Observatory)

which the sunrise oscillates. Midway between is equinox sunrise. For convenience, additional sunrises, midway between each solstice and the subsequent equinox, might be noted. Alexander Thom has found alignments of British megaliths marking these solar declinations and offers further evidence for division of the year into sixteen intervals. Professor Thom has carefully measured, mapped, and analyzed over three hundred Megalithic sites in Britain and France. Almost singlehandedly he has established the standards for archaeoastronomical fieldwork and interpretation, and his amazing results have stirred controversy during the last three decades. According to Thom, a position near the equinox, and not precisely on it, was used to counteract the effect of the ellipticity of the earth's orbit on the calendar.

Stonehenge, it has long been said, includes an alignment, from the

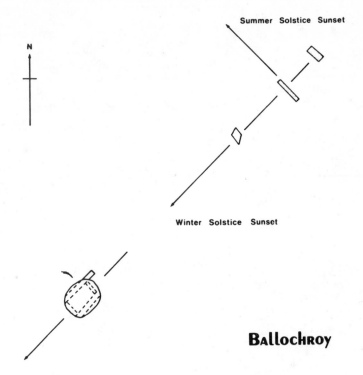

BallochRoy

Ballochroy, on Scotland's Kintyre peninsula, is an alignment of three stones and a kist, or stone chamber. The line to the southwest points to a feature on the small island of Cara and the point of winter solstice sunset. The center stone appears to indicate the foresight to the northwest for the summer solstice sunset. (Griffith Observatory, after Alexander Thom)

monument's center to the tip of the Heel Stone, directed toward the summer solstice sunrise. Many precise solstice alignments have been found among the megaliths of Britain and France by Thom. One of the best of these is at Ballochroy, on Scotland's Kintyre peninsula, overlooking the Sound of Jura. Kintraw, in Argyllshire and about forty miles, by road, north of Ballochroy, is another well-substantiated solstice indicator. Sir Norman Lockyer claimed that the Great Temple of Amen-Ra at Karnak, Egypt, was aligned on the summer solstice sunset. In recent years, Gerald Hawkins has supported a solstitial interpretation of this temple, but he has turned it around 180 degrees to the winter solstice sunrise. Dr. John Eddy examined the Big Horn Medicine Wheel in Wyoming and demonstrated its solstitial character. At Uaxactún in Guatemala, the Maya may have constructed a set of solstice oriented platforms.

Sunrise on the summer solstice at the Big Horn Medicine Wheel, in Wyoming, appears over the central cairn when viewed from the outlying cairn southwest of the wheel's rim. (John A. Eddy)

THE MYSTERIOUS MOON

Despite the many examples of ancient and prehistoric solar observatories, there are many monuments that do not fit the pattern. There are calendars in use, too, that are not calibrated by the sun. This should come as no surprise, however, for the moon's rapid and obvious changes also provide a ready means for keeping track of the date, and a lunar calendar is still used in the Islamic world. A lunar calendar is serviceable enough for day-to-day activities, but it has the disadvantage of getting out of step with the seasons. Reconciliation of the lunar and solar calendars required much of the energy expended by the calendar reformers of Western Europe. Of course, if the solar year contained an exact number of lunations, or lunar cycles, the two calendars would remain in phase. For better or worse, they do not, and lunar calendars are often restricted to religious use. The date of Easter is still determined by the moon. This means that on our civil, solar calendar the date of Easter can wander several weeks.

Each daily rotation of the earth brings the moon into view over the eastern horizon and sets it down below the west. This pattern would be similar to the sun's, but the moon is in orbit around the earth. From the earth the moon appears to move eastward through the stars, about 12½ degrees each day. Because the apparent motion of the moon is eastward, the earth must rotate 12½ degrees further to the east each day to again bring the moon into view. The earth requires approximately fifty minutes to rotate through the additional angle, and therefore moonrise is delayed about fifty minutes from one day to the next.

As the moon moves in its orbit about the earth, the moon's position with respect to the sun and the earth changes. Shining only by the reflected light of the sun, the moon alters its appearance. Only one half of the moon is illuminated at one time; the rest is shadow. Through the course of the month the half of the moon that faces earth cycles from complete darkness to complete illumination and back to darkness again. The moon is said to be full when the earth-facing side is fully lighted. The time from full moon to full moon is the synodic month, and it lasts about twenty-nine and a half days. The word "synodic" derives from a root word which means a "coming together," or conjunction. In astronomy, "conjunction" refers to a close configuration in the sky of two objects, but the term "synodic" has been extended to successive corresponding phases of the moon.

The cycle of the lunar phases is simple and familiar. During the two days or so that the moon is in conjunction with the sun, the moon's lighted half faces away from earth, and we see no moon, or a new moon. As the illuminated portion of the moon shifts into view, the moon appears to grow brighter and more complete from night to night. During this period the moon is said to be "waxing." First we see a thin crescent that is most noticeable in the early evening. It hangs like the smile of the Cheshire Cat over the western horizon, and it sets a while after the sun.

About one quarter of the way through the synodic month, the angle between the moon, earth, and sun is 90 degrees, and a so-called half moon is visible. This phase is called "first quarter," however, in reference to the progress of the monthly cycle. The new moon, in conjunction with the sun, rose and set with the sun. A first-quarter moon is 90 degrees from the sun, and therefore this moon rises about six hours after sunrise and sets about six hours after sunset.

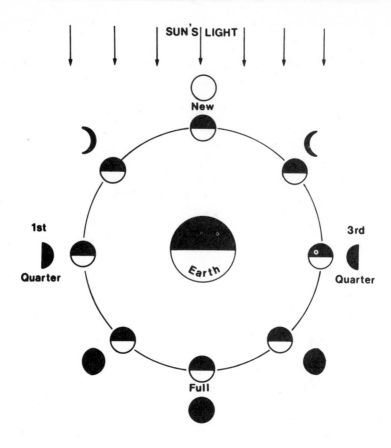

In a single "moonth," or month, the moon orbits the earth and passes through a complete sequence of phases. The word "moon" is related to an ancient word which means "to measure," and the pattern of lunar phases may have been one of the first cycles used by people to measure the passage of time. (Griffith Observatory)

The moon continues to wax through a near full, or gibbous, phase. Both sides of the moon's edge, or limb, are convex. This bulging appearance explains the term "gibbous," meaning "humped." Full moon occurs when the moon is opposite the sun. Therefore the full moon rises when the sun sets, and moonset occurs at sunrise.

During the second half of the month the phase of the moon progresses through a symmetric set of phases: gibbous, half, crescent, and back to new. The moon is said to be "waning" during this period of ever diminishing brightness. The half-moon phase of the waning moon finds the other half of the earth-facing side of the moon illuminated. This phase marks the end of the third quarter and the beginning of the last quarter of the synodic month. It is called "third" or "last quarter." The third-quarter moon is 90 degrees from the sun but on the other side of the moon's orbit from first quarter. Therefore the third-quarter moon rises about six hours after sunset and sets about six hours after sunrise. The relationship between the phase of the moon and the position of the sun permits us to tell time by simply observing the phase of the moon and its angle in the sky. We count time by the sun's position, but the moon can be used to infer the sun's position and would be useful as a clock, especially when it was visible during the night.

Not only do the moon's phase and its times of rising and setting change each day, but the horizon positions of moonrise and moonset also change. If the moon orbited directly above the earth's equator so that the moon's path would coincide with the celestial equator, no changes in the position of moonrise or moonset would be observed. Instead, the moon's path falls close to the ecliptic, and so its rising and setting positions roughly coincide with those of the sun, though not at the same place at the same time. The moonrise oscillates between an extreme northeast point to an extreme southeast point and back to the northeast again, but the moon goes through this cycle in a month rather than in a year. Because the full moon rises opposite the sun, the full moon that occurs near the time of the summer solstice will rise in the southeast, opposite the northwest-setting sun. In winter the sun sets in the southwest, and so the full moon rises in the northeast, near the summer solstice sunrise point. Through the course of a single month the moonrise at new moon coincides with the sunrise. At full moon, moonrise is opposite the sunset. For first and last quarter, the moonrise occurs in between. The pattern follows the seasons as tabulated below.

MOONRISE POSITION

	summer	fall	winter	spring
New	northeast	east	southeast	east
First quarter	east	southeast	east	northeast
Full	southeast	east	northeast	east
Third quarter	east	northeast	east	southeast

If the moon's path coincided exactly with the ecliptic, the full moon would rise at the same position at the same time of the year every year. Different phases would occupy that same moonrise point at other times of the year every year. Furthermore, an eclipse of the sun would occur each month at new moon. We would likewise expect to see an eclipse of the moon each month at full moon. Of course, eclipses don't occur that frequently, and the reason is the inclination, or tilt, of the moon's orbit. The moon's orbit is inclined approximately 5 degrees to the earth's orbit around the sun.

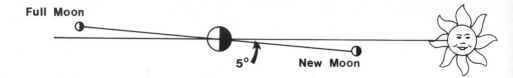

The moon's orbit around the earth is tilted 5 degrees to the earth's orbit around the sun. For this reason the moon, the earth, and the sun do not always fall on the same line at every new and full moon. (Griffith Observatory)

The inclination of the moon's orbit might be expected to oscillate the moon to an angular extreme above and below the celestial equator each month, just as inclination of the ecliptic carries the sun above and below the celestial equator each year. The exact value of the extreme positions, or declinations, would depend on the orientation of the moon's orbit to the earth's orbit. If it is imagined that the maximum extreme positions are reached in the summer, with the new moon at a declination of 28½ degrees (23½+5) above the celestial equator, the full moon of that same month would be found at 28½ degrees below the celestial equator. On the horizon, moonrise at new moon would be positioned a little north of the summer solstice sunrise point. Moonrise at full moon would fall a bit to the south of the winter solstice sunrise point. Eclipses would occur only at the new and full moons that fall on the ecliptic and, therefore, midway between the two extremes. In this situation eclipses would be expected in spring and fall.

We can as easily imagine the moon's orbit to be turned so that the moon appears below the ecliptic when the sun and the ecliptic are highest in the sky, in summer. Under these circumstances the new moon will rise a bit south of the summer solstice sunrise point, and its declination will be 18½ degrees (23½−5). Similarly, the full moon would rise a bit north of the winter solstice sunrise point in the southeast and would have a declination of −18½ degrees.

As noted above, the exact values of the positions of the lunar extremes depend on exactly how the tilted orbit of the moon is oriented. Assuming the orientation held constant, we would expect the moon to reach the same extremes each month, and, furthermore, we would expect to see the full cycle of the moon's phases at that extreme during the course of the year. The full moon would be seen at the extreme north position once during the year, in winter, and the other phases would be seen there throughout the rest of the year. The situation is not so simple, however. The moon's orbit precesses, or turns, with respect to the earth's orbit. The moon's path in the sky therefore varies with respect to the equator and ecliptic.

The moon's orbit revolves from east to west. The motion of the moon in its orbit is from west to east. The precession of the moon's orbit is in the opposite sense and is therefore called a regression. The intersections of the moon's orbit with the ecliptic are called the "nodes" of the moon's orbit, and a line drawn to connect the two nodes is called the "line of nodes." The backward precession of the moon's orbit is therefore called the "regression of the line of nodes." This line can be imagined to turn in a counterclockwise direction when the orbit is viewed from the north celestial sphere. Regression of the line of nodes is very important information, for eclipses can only take place when the centers of the sun, moon, and earth all fall along the same line. This is equivalent to the requirement that the new and full moons occupy either of the nodes when the sun does, or that the sun, earth, and moon all fall along the line of nodes.

A solar eclipse is caused by the passing of the moon directly in front of the sun to obscure the sun's light. Although the sun is about four hundred times larger than the moon, the sun, by coincidence, is also about four hundred times farther away. The sun and the moon appear to be roughly the same size in the sky (about ½ degree), and the moon is able to eclipse the sun, provided it coincides with the sun's position in the sky. This can only happen where the sun's path and moon's path coincide, on a node.

Lunar eclipses occur when the moon passes into the earth's shadow. The earth's deep shadow, or umbra, is always present and extends approximately 855,000 miles into space. The shadow is cone-shaped and at the distance of the moon's orbit (only thirty earth diameters away) is about 5,700 miles across. The moon's diameter is about 2,160 miles. The shadow at the moon's orbit is about two and a half times larger than the moon, and if the moon falls within the shadow, it will be eclipsed. The earth's shadow is centered on the line that connects the earth and the sun. A lunar eclipse can occur only when the shadow also falls upon the moon. For this to happen, the sun and moon must occupy opposite nodes of the moon's orbit.

The earth perturbs the moon in its orbit. These gravitational forces cause the line of nodes to precess, and a complete regression takes 18.61 years to complete. Over an 18.61-year period the northern moonrise extreme moves from a maximum to a minimum and back to a maximum. The same is true for the southern moonrise. The period of maximum extreme has been called the "major standstill" by Alexander Thom; the minimum corresponds to his "minor standstill." The moon does not really stand still, but it reaches close to the extreme declination with little noticeable change from month to month for several months. In that way a standstill is analogous to a solstice, or "sun standstill," but the

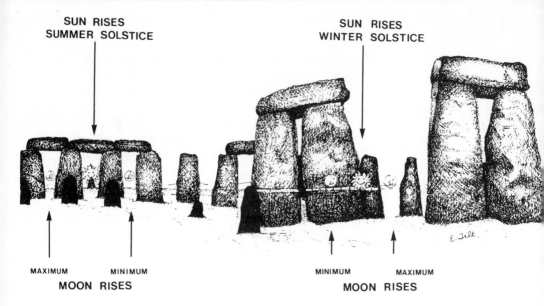

SUN RISES
SUMMER SOLSTICE

SUN RISES
WINTER SOLSTICE

MAXIMUM   MINIMUM

MOON RISES

MINIMUM   MAXIMUM

MOON RISES

The standstill limits of the moonrise and the solstice limits of the sunrise as seen from the center of Stonehenge indicate that considerable movement of the sun and moon could be observed on the horizon over the course of their cycles. Stones block the view of some of the southern risings, but these were probably not intended to be seen from the monument's center. (Griffith Observatory)

A lunar eclipse may occur when the moon crosses the ecliptic. If the sun is on one node and the moon on another, the earth, moon, and sun fall on the same line—the line of nodes—and the moon enters the earth's shadow and is eclipsed. When the moon is exactly between the earth and the sun, its shadow may just reach the earth to produce a solar eclipse. (Griffith Observatory)

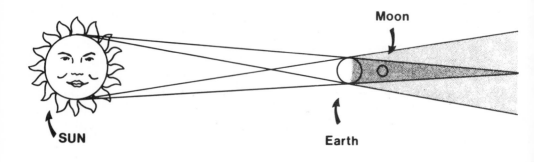

Moon

SUN

Earth

parallel doesn't really hold. In the year of a major standstill, the winter full moonrise will reach its northernmost position. In the same year the summer full moonrise reaches its southernmost extreme. In 9.3 years the minor standstill occurs. Now the winter full moon still rises in the northeast but as far south of the summer solstice sunrise point as it ever reaches. In the summer of the same year the full moon still rises in the southeast but as far north of the winter solstice sunrise point as it will reach that cycle. Although this pattern is virtually correct, it should be remembered that the period of regression is not an even multiple of tropical years but eighteen and a fraction. This means that the standstills will not occur at exactly the same time of year each year they occur.

The changing position of moonrise is also accompanied by a change in the height of the moon as it crosses the sky. This effect is particularly noticeable for the full moon, whose light is so important to a culture lacking electric illumination. Over the 18.61 year cycle, the difference in the height of the full moon and the difference in the duration of its appearance above the horizon may well have been of concern to prehistoric and ancient peoples, especially those in the more northern latitudes where the effect is more pronounced.

One node of the moon's orbit marks the moon's passage from below the ecliptic to above it and is called the "ascending node." The opposite node is naturally the "descending node." It takes 346.62 days, or one eclipse year, for the sun to travel from the ascending node to the descending node and back again to the start. An eclipse year is approximately twenty days shorter than a tropical year because the ascending node regresses eastward to meet the sun about twenty days earlier each year. If we characterize the sun's occupation of a node as an eclipse season, eclipse seasons occur twice a year, or once every 173.31 days. They each arrive about twenty days earlier than the year before.

An interval of 27.21 days is required for the moon to complete the circuit from one coincidence with the ascending node to the next. This period is called a "draconic month," and it is shorter than the synodic month by a little over two days. Most of this is a result of the earth's motion in orbit about the sun, but a small part of the difference is generated by the eastward regression of the ascending node to meet the moon before it reaches the same phase it displayed while at the node the previous month.

Complicating the already complex lunar motion is a small wobble that causes the moon's declination to oscillate plus or minus nine arc minutes on top of its expected cyclical variations. The slight oscillation is called the "inclination perturbation." It is caused by the tilt of the moon's orbit and the difference in the sun's gravitational force on

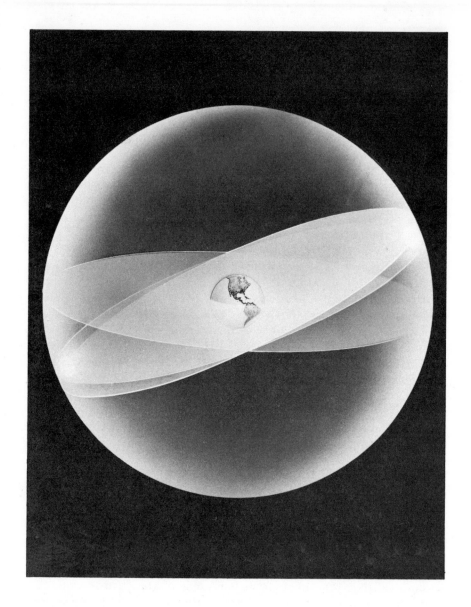

At the time of the major standstill, the moon's orbit carries the moon well above and well below the ecliptic in a single month. In this picture the disc of the moon is shown at both its extremes, above and below the ecliptic. (Griffith Observatory, John Lubs)

The minor standstill occurs 9.3 years after the major standstill. Now for several months the moon's orbit allows the moon to swing each month between limits that are well inside the extremes of the ecliptic, that is, between the solstices. Again, the moon is shown on opposite sides of its path. (Griffith Observatory, John Lubs)

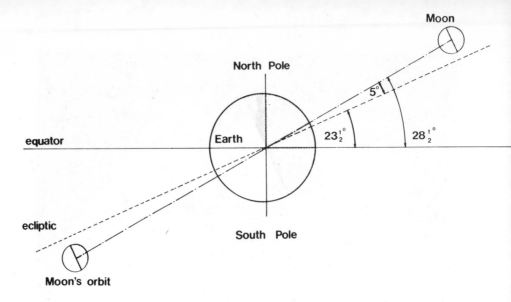

Major Standstill at Winter Solstice

At the time of winter solstice the sun shines more directly on the earth's southern hemisphere—in this diagram, from the left. At summer solstice the sun would shine from the upper right. The line from the sun therefore represents the ecliptic. The earth's equator makes an angle of 23½ degrees to the ecliptic, and here, at the time of major standstill, the moon's orbit is set 28½ degrees to the equator. New moon occurs when the moon is roughly in the same direction as the sun 28½ degrees below the celestial equator. Full moon, on the other hand, finds the moon opposite the sun and 28½ degrees above the equator. (Griffith Observatory)

The minor standstill will find the moon's orbit now set 18½ degrees (23½−5) to the celestial equator. Here at winter solstice again, the sun shines from the lower left and parallel to the ecliptic. The new moon, again below the celestial equator, will reach a position 10 degrees closer to the equator than at major standstill. Similarly, full moon at maximum height above the equator will be 10 degrees lower than 9.3 years earlier. (Griffith Observatory)

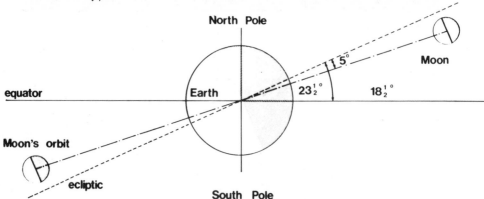

Minor Standstill at Winter Solstice

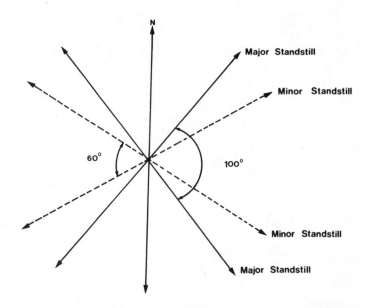

Extreme Azimuths Of The Moon At Stonehenge

During the 18.61-year lunar standstill cycle, the separation between northern and southern moonrises is observed from the latitude of Stonehenge to vary from 100 degrees at the major standstill to 60 degrees at the minor standstill, 9.3 years later. These shifts in moonrises and moonsets are large and dramatic. (Griffith Observatory)

The path of the full moon during the nights near the winter solstice carries the moon from its northeastern rising point high overhead and allows it to set in the northwest. The full moon rises at sunset and sets at sunrise, and, conveniently, the moon shines throughout the long winter night. In summer the full moon rises in the southeast, transits low across the southern sky, and sets in the southwest. The night is short, as is the path of the moon across the sky. In this illustration, at the major standstill, the moon's monthly northern extremes are as far north as they are ever observed. Similarly, the southern extremes are at their greatest limit. (Griffith Observatory, John Lubs)

At minor standstill the moon's behavior is basically the same. Now, how-ever, the winter full moon does not rise and set so far to the north, nor does the summer full moon reach so far south. The moon is never seen inside these inside limits of its risings and settings during the course of the entire 18.61-year cycle. (Griffith Observatory, John Lubs)

either side of the orbit. When the moon is on the solar side of its orbit, it experiences a different force than when it is on the far side of its orbit. The inclination perturbation is at its maximum positive value when the sun is at a node. During the next 173.3 days, or half an eclipse year, the sun moves to the other node. Over the same interval the inclination perturbation has gradually diminished to zero, dropped to its maximum negative displacement, increased back to zero, and once again achieved its maximum positive shift.

To see how all of the motions of the moon combine to allow it to reach a maximum declination, north or south, we must consider the standstills. A standstill occurs when the nodes coincide with the equinoxes. At major standstill the ascending node and the vernal equinox are near each other. After 9.3 years the descending node and the vernal equinox will coincide, and it will take another 9.3 years to go from this minor standstill back to the major standstill. The maximum effect of the inclination perturbation will be visible when the sun is at either node. At major standstill this occurs at the time of vernal equinox and at the time of autumnal equinox. At the time of vernal equinox the moon occupies its greatest declinations north and south at first and third quarter respectively. The situation is reversed at the time of autumnal equinox. To observe the moon on the horizon at night in spring, we must observe the setting first-quarter moon to obtain the northernmost position. In autumn the rising third-quarter moon provides the necessary information.

The subtle inclination perturbation was discovered in the sixteenth century by Tycho Brahe, the Danish astronomer, but it may have been suspected by tenth-century Arab astronomers. By contrast, the 18.61-year cycle of standstills has such an obvious effect on the moonrises, moonsets, and the moon's time above the horizon that it was probably known from prehistoric times. Alexander Thom offers evidence that the inclination perturbation was also observed by Neolithic astronomers.

Many small variations in the motion of the moon and the earth alter the exact relationship between the earth, moon, and sun when the sun and moon nearly occupy the nodes. Eclipses can occur only when the moon and the sun are within certain angular limits of the nodes. These ecliptic limits are determined by the apparent sizes of the moon and the sun and their distances from the earth. The ellipticity of the earth's and moon's orbits and the precession of the moon's line of nodes all complicate the timing of an eclipse. Furthermore, the area on earth over which an eclipse may be seen, when it does occur, is limited for a lunar eclipse and even narrower for a solar eclipse. All this makes the prediction of eclipses a difficult and tricky business. Even a successful method may not be confirmable. It is possible through accumulation of records, nevertheless, to construct a cycle of eclipses which will help predict future eclipses. One such cycle is called the "saros." The saros is a sequence of eclipses for which similar conditions repeat, about every eighteen years. Although the saros can be used to anticipate an impending eclipse, it does not permit an infallible prediction.

Without a written language a culture would find the saros and other eclipse cycles difficult to apply. It may be possible to predict eclipses through direct observation of the moon's position, if it were known that the maximum positive displacement of the inclination pertur-bation occurs at the eclipse seasons. The perturbation cycle is going on all of the time, but the displacement is normally mixed in with the other contributions to the moon's declination. The perturbation maxi-mum can only be detected at one of the standstills, when the perturba-tion shifts the moonrise and moonset slightly beyond the expected position on the horizon. Once the observation is made, a Megalithic astronomer need only count days until half an eclipse year has gone by. Since the perturbation maximum can be observed only at the stand-stills, the prehistoric astronomers could not afford to miss them. As it is, the cycle can only be calibrated every 9.3 years, or possibly when an eclipse is observed. Alexander Thom has assembled considerable evi-dence that the Megalithic people built Stonehenge and other sites, in-cluding Temple Wood in Argyll, in Scotland, as precise lunar observa-tories where genuine, practical astronomy was carried out.

THE SUBTLE STARS

Observations of stars could have been made easily by prehistoric astron-omers. The daily rotation of the earth would have been reflected not only in the sun's motion across the sky in the daytime but also in the risings and settings of stars at night. The earth's annual orbiting of the sun makes itself felt by changes in the rising and setting points of the sun throughout the year in the duration of daylight. The sun is also moving through the constellations of the background stars, but because the sun is so bright, we cannot see this motion directly. Astronomers in prehistory would have noticed, however, that some groups of stars domi-nated the winter nights and others the summer.

A calendar could be calibrated by the first or last appearance of a group like the Pleiades, or Seven Sisters. It is said that the Celtic peo-ples attached special significance to the Pleiades at their dawn rising, which coincided with Beltane, or May Day. The Aztecs measured the completion of one fifty-two-year cycle and the beginning of the next by the date of culmination of the Pleiades at midnight. Teotihuacán, in central Mexico, antedates the Aztecs by over a thousand years, yet the entire plan of this huge metropolis was based upon the direction of the setting of the Pleiades.

Entire sequences of stars were used by the Egyptians to tell time at night, and their choice of stars and calendar calibrator was responsible for our present convention of twenty-four hours in the day. One more example of this kind of thing is mentioned by Alexander Thom, who suggested that the Megalith Builders also used certain bright stars as nighttime clocks.

Stars were used to calibrate the solar calendar and to mark the progress of the seasons. One of the best stellar phenomena for these purposes is the heliacal rising. A star rises heliacally when it is first visible in the dawn sky, before sunrise. There is a period during the year when a particular star rises in the nighttime. Gradually the sun catches up to that star. The sun's motion along the ecliptic carries it around the sky. For a period of a few months the star rises in the daytime, when the sun is up. The star is invisible. Eventually the sun moves far enough east of the star to permit the star to rise before the sun. The star can then be seen again, and its first dawn appearance is its heliacal rising.

The heliacal rising of Sirius, the Dog Star, was extremely important to the Egyptians. During one period of their history, the heliacal rising of Sirius occurred at the same time as the summer solstice and, by coincidence, at the same time of the Nile inundation. The Nile and the agricultural cycle dominated Egyptian life. The Egyptians based their calendar upon them. The calendar was calibrated, and the year began, with the heliacal rising of Sirius. John Eddy interprets alignments at the Big Horn Medicine Wheel in Wyoming as indicators of a sequence of heliacally rising, bright stars which warned the Plains Indians who used this site of the impending cold weather.

Most cultures have synthesized a celestial geography from the patterns of stars of the night sky. Groups of stars are associated together to form constellations. The constellations often represent or symbolize important elements of the culture's mythology. We do not know for certain who invented the constellations still in use today. Certainly many were in use as early as 500 B.C. in Babylonia, but their antiquity may be greater.

Unlike the sun and the moon, stars do not visibly move with respect to one another. Year after year the cycle is repeated. However, over several centuries another motion of the earth, called "precession," will gradually affect the seemingly immutable stars. Precession is a wobble of the earth's polar axis. The wobble is caused primarily by the gravitational pull of the sun and the moon on the earth's equatorial bulge. Instead of spinning with its axis forever fixed in one direction, the earth wobbles around like a top spinning on the floor. The top's period of precession is a few moments. For the earth twenty-six thousand years are needed. During this period the north pole swings around the sky, sometimes pointing toward our north star, Polaris, and sometimes toward another star. The equinoxes and the solstices shift with respect to the background stars. Stars of the summer sky are later seen in winter and still later, at the cycle's end, in summer again. A star that lines up with some monument at a certain epoch will not line up with that monument several centuries later. The motion is subtle, but it can be detected if observations are made over a long period. The earliest known direct reference to precession is that of the Greek astronomer Hipparchus (second century B.C.), who is credited with discovering it. Adjustments of Egyptian temple alignments, pointed out by Sir Norman Lockyer, may well indicate a much earlier sensitivity to this phenomenon, however. Recently Giorgio De Santillana and Hertha von Dechend, both historians of science, in *Hamlet's Mill* (1969), have argued that much mythological narrative is really a symbolic description of the earth's precession as viewed in the stars.

IN SEARCH OF ANCIENT ASTRONOMIES

Evidence of the astronomical orientation of monuments is found in many places, but we still do not know how long humans have been making astronomical observations. It is often thought that scientific astronomy began with the Greeks, but this is simply not so. The Babylonians developed a sophisticated system of keeping records and methods of calculation that indicate careful observation of the sun, moon, and planets. It is known that they attempted to predict eclipses. But astronomical observation is also possible in preliterate societies. We have no idea how long people, or creatures resembling people, have been watching the sky. We are not certain when a sense of time first became part of the human consciousness. It is not difficult to imagine, with the evidence of human fossils dated at 4 million years, that we have been noticing the sky for some time. Tallies of the moon's phases may have been made in the Paleolithic Age. Even today we find aborigines using a gnomon to determine time and seasons astronomically.

Alexander Marshack, formerly research fellow at the Peabody Museum of Archaeology and Ethnology, in *The Roots of Civilization* (1972), proposed that Paleolithic notations on bones correspond to lunar observations. Other suggestive markings on cave walls evoke the same suspicion. Some of the bones Marshack has analyzed are in-

The time is late spring in ancient Egypt. Sirius has risen in the east, across the Nile from the pyramids, but the sun is already up. Sirius, therefore, cannot be seen yet when it rises. (Griffith Observatory)

The motion of the earth in its orbit around the sun makes the sun appear to move east with respect to the stars. A few weeks before the summer solstice the sun rises when Sirius rises, but the glare of the dawn is still too bright to allow Sirius to be seen. (Griffith Observatory)

Finally, the sun has moved far enough to the east of Sirius to permit Sirius to be seen, for the first time this year, in the predawn sky. By coincidence it is the summer solstice. This heliacal rising of Sirius roughly coincided with the flooding of the Nile, and for the Egyptians this association of star, river, and summer solstice was of great importance. (Griffith Observatory)

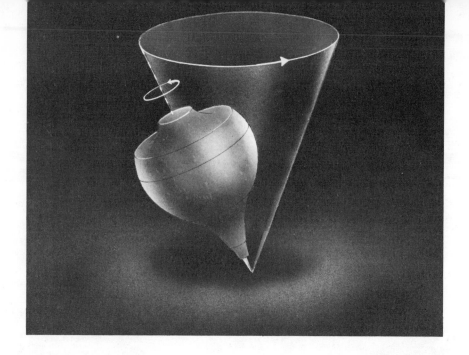

Precession of the earth is similar to the wobbling of a spinning top. Gravity tries to pull the top down to the floor, but because the top is spinning, this force makes the top swivel as it spins. (Griffith Observatory, John Lubs)

The gravitational attraction of the moon and sun acts to straighten the earth's axis upright with respect to the ecliptic. The earth is spinning, however, and so these forces cause the earth to precess, or swivel, with a period of twenty-six thousand years. The axis of the earth now points toward Polaris along the dashed line pointing to the upper left, but precession will shift this axis toward the star Vega twelve thousand years from now along the line pointing to the upper right. This shift of the position of pole in turn shifts the rising and setting points of stars, but the effect takes a few centuries to become noticeable. (Griffith Observatory, John Lubs)

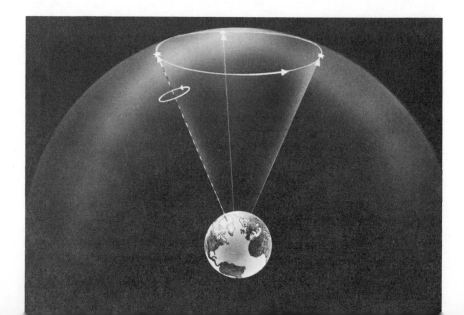

scribed with simple sequences of lines. The sequences are spaced, and from one set to the next the style may change. The lines may be vertical in one set and slanted in the next. Some bones, are more complex. The Blanchard bone, a small piece of bone found in the Dordogne region of France inscribed by some Cro-Magnon individual about twenty thousand years ago, has a complicated pattern of marks. The shapes of the marks vary, and the sequence curves around in a serpentine pattern. In Marshack's view the turns in the sequence represent, on one side, the times of dark, new moon, and on the other, bright, full moon. Statistical analyses may not support Marshack's interpretations, but similar batons and sticks are carved for the same purpose by the Nicobar Islanders in the Bay of Bengal.

The impact of the seasons on human activity is evident. The keeping of the calendar and the telling of time probably prompted, at some time in prehistory, observation of the moon and sun. Or perhaps it was the other way around. We may never be able to know how people at this stage of cultural development viewed the world around them. But we can guess. The sun, as seen from the earth, is simultaneously the cause of the seasons and the indicator of their passing. To understand, even superficially, how early people might have regarded the sun, we must integrate these two roles of the sun into a single concept. Perhaps in this way we can incorporate ourselves into the prehistoric landscape. By unifying the immediate experience of a natural event with its symbolic meaning, we may at least get some imperfect idea of how prehistoric people regarded and approached the world around them. True sun worship would amount to a sensible recognition of the sun's importance to cycles of life on earth and practical observation of the sun's behavior, which reveals the pattern of its effects. The moon and the stars may well have been understood in similar terms.

Perhaps the urge to orient the landscape of space and time is a fundamental, practical response of the human brain, an attribute of our minds that permits them to function at all in the chaos of events and objects that make up the universe around us. This urge may be what prompted us to observe the sky and become astronomers in the first place. Our ancestors may have pulled patterns from the sky and incorporated them into their architecture.

More often than not we have only the ruins to guide us. If we can extract from the European megaliths, the New World ceremonial centers, or the Egyptian temples evidence of observation of the kinds of obvious and important astronomical phenomena discussed here, we may gain some insight into the needs and the evolution of the human mind.

CHAPTER TWO

Rings and Menhirs: Geometry and Astronomy in the Neolithic Age

ALEXANDER THOM AND ARCHIBALD STEVENSON THOM

Of all the prehistoric monuments and ancient ruins around the world Stonehenge, in England, is certainly one of the most celebrated. Few are aware, however, that Stonehenge shares the British Isles with over nine hundred other stone rings, and these are just the ones that are left. Together with thousands of individual standing stones, barrows, cairns, and other Megalithic antiquities, they indicate the great energy and mysterious motivation of the Neolithic and Bronze Age people who built them. These Megalith Builders, working in earth and stone, left these remnants of their presence throughout Northern and Western Europe over many centuries, from 4000 B.C. to 1500 B.C.

It has been fashionable to interpret the design of Stonehenge astronomically, but it is not so well known that an even stronger case has been made by Professor Alexander Thom for many of the other Megalithic sites. Thom has personally surveyed over three hundred of the old sites. He is the author of two books, Megalithic Sites in Britain *and* Megalithic Lunar Observatories, *and dozens of scholarly articles. His work over the last fifty years has made him the acknowledged leader in archaeoastronomy. The great accuracy of his field measurements, the high precision of his analysis, the volume of his work, and his carefully considered yet revolutionary results have set the standards for this young science.*

Thom, emeritus professor of engineering science at Oxford, is now in his eighty-fourth year and is still very active. R. J. C. Atkinson, the foremost authority on Stonehenge, has added that Thom's more recent work among the thousands of stones at Carnac in Brittany is "one of the most notable and difficult combined operations in prehistoric research to have been mounted anywhere in Europe in the present century."

"Megalith" means "great stone" (mega, great; lith, stone), and Thom has shown that these great stones have mathematical significance, in addition to their use for observation of the sun and moon. Stone rings in a variety of shapes—circles, flattened circles, egg-shaped rings, and ellipses—all show an undreamed of precision and complexity of design.

Professor Thom's son, Dr. A. S. Thom, is senior lecturer in engineering at Glasgow University. Dr. Thom has also participated in many surveys and has coauthored studies of Carnac in France, Brogar in the Orkney Islands, and Avebury and Stonehenge in England. He has also lectured internationally on these results. Together, the Thoms here describe some of the many ruins they have examined and report on the geometry, systems of measurement, and astronomy practiced by the Megalithic peoples four thousand years ago. Some of their conclusions are extraordinary, for they imply that the prehistoric rules of mathematics and astronomy developed without the benefit of a written language. If this is true, our understanding of how our culture came to be what it is will have to be revised.

We write this survey of Megalithic geometry and astronomy in English. To do so we must have a working knowledge of grammar and syntax. That is, we must know the rules.

Our Highland friends tell us that Gaelic is a much more powerful and perfect language than English. Doubtless the Welsh would say the same for their mother tongue. Classicists maintain that the most perfect language of all is probably Ancient Greek, which had a much more complex set of rules. Certainly mastery of all of the complications of written Greek is difficult.

Greek existed three thousand years ago. From where did it come? How long did it take to develop? If one thousand babies were left on an island and were not allowed to hear any language at all, how long would it take for a working language to develop? How many generations would pass before a really first-class language, with grammar, syntax, and vocabulary, appeared: one generation? ten? a thousand? We,

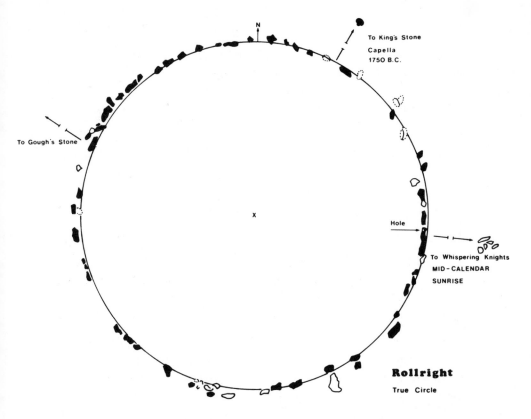

We know the Megalith Builders could lay out a good circle when they wanted to. Rollright, in Oxfordshire, is a true circle with a diameter of approximately 38 Megalithic yards. (Griffith Observatory, after Alexander Thom)

of course, cannot answer the question, for we do not really know how human language actually evolved. And once a language exists, it continues to evolve. Rules are discarded. The language may deteriorate and become ineffective. If it is lost, much of the culture of the people who used it may also be lost.

A developing civilization must have a working language. It also needs a working knowledge of mathematics, and here many of the same things apply. A set of rules must be developed. Probably a start is made with a false set, but mathematics is itself capable of showing up the falsehoods.

Our purpose here is to examine the rules of early mathematics. We can discover some of these in, for example, petroglyphs carved on rocks. Mathematical designs were also executed in huge stone rings. One of the most complicated patterns is found at Avebury, in Wiltshire, England, just seventeen miles north of Stonehenge. It might seem at first that the large ring at Avebury is just a poorly drawn circle, but even at Avebury remnants of carefully laid out circles are found inside the earthwork enclosure and the large compound ring. The geometry has been deciphered, and its agreement with the actual accurate survey forces us to conclude that the large ring is a result of intentional design.

MEGALITHIC MEASUREMENT

To discover the elementary rules of Megalithic geometry, we must first investigate the simpler designs of the many circles scattered over the moors of Britain. Probably many of these circles of stones at the first were set out with a rope, but the more precise designs must have been set out on the ground with a standard length of measurement. Careful analysis of several hundred sites indicates that this length was 2.72 feet. This unit is sufficiently close to 3 feet to call it, for convenience, the "Megalithic yard." The word "yard" means a piece of wood, a rod, a pole, or a stick. The French *verge* means the same thing, and, again, the Spanish *vara* also means a length of wood. Curiously, the old Spanish *vara* was almost the same length as a Megalithic yard.

Statisticians have spent considerable time analyzing the available diameters of carefully measured rings and admit that, in the Scottish circles at least, there exists definite evidence for the Megalithic yard. They also say that the English circles do not provide such definitive evidence, but this is because the English data include the more complicated designs which do not lend themselves well to statistical analysis.

Several factors make it difficult to obtain the exact original design of any Megalithic ring. Many of the sites in Great Britain have been badly disturbed, and in recent years some of them have been equally badly restored. Many of the originally upright stones at Rough Tor, in Cornwall, for instance, are now fallen. Some of those remaining uprights do not adhere closely to the flattened circle design. Flow of the soil or earth movement due to frost may be responsible.

A boulder clay on a slope of 1 degree tends to flow. If the underlying rock is irregularly distributed, the surface movement will be irregular. A deeply embedded stone will move differently from one whose foundation is in the surface. A tree can grow around a stone, lift it from the ring, and then rot away. Or the tree can simply press a stone on one side. In a hard frost the sun warms the south side of a stone and softens the ground on that side. Later, the freeze may return, and the stone will be minutely disturbed. Several thousand years of this activity can make the displacement considerable. All of these factors can disguise the original plan of those who set out the stone rings.

From a practical point of view it would have been very sensible for the Megalith Builders to have used a standard unit of length. The development of a system of units of measurement is a fundamental step in the evolution of architecture, and a building or construction of any sort can hardly be engineered and built without some means of setting out its dimensions.

Even if we assume that the Megalith Builders did set out their rings of stones in some standard length, say the Megalithic yard, we don't necessarily know how the dimensions of the rings were actually measured. Did the builders measure a diameter to the inside surfaces, to the outside surfaces, or to the centers of the stones? It seems likely that certain dimensions, say, the diameters or the circumferences, were chosen to be whole, or integral, multiples of the standard unit. For example, the diameter of a circle might be set out as 8 Megalithic yards exactly, not 7.85 nor 8.25. A decision like this may have been made to satisfy some principle of design. Probably many of the other dimensions of the structure would not be integers at all; other factors might control other components of the layout. The Megalith Builders left no written records and so we do not know, as we begin a search for their units of measurement, what principles of design most satisfied them. Nor do we know what other considerations were incorporated into their monuments. All of these uncertainties complicate analysis of the stone rings. Nevertheless, it has been possible to deduce the units in which the monuments were built from the monuments themselves, and, in turn, some of the possible principles of design have also become evident.

All of our measurements of stone rings made up to 1967 are shown in a diagram which reveals that certain diameters are whole multiples of the Megalithic yard (MY) and were favored dimensions of the builders. Particularly high peaks occur at significant circumferences also, namely at 25 MY, 50 MY, and so on. The builders of the circles also evidently wanted the circumferences to be integral multiples of the unit. A circle whose diameter is eight units has a circumference of almost exactly twenty-five units. Indeed, if the ratio between circumference and diameter is thought to be 3⅛ (instead of 3⅐, a closer approximation of π), a diameter of eight units gives a circumference of exactly twenty-five units. The discovery of the existence of certain circles whose integral diameters yield integral circumferences may have led to the development of a rule of Megalithic mathematics and design: the circumference must be in whole units of Megalithic rods (MR), where an MR is defined to be 2½ MY. This rule means that a circle whose diameter is 8 MY has a circumference of 10 MR.

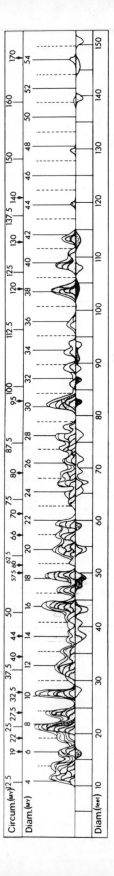

A graphical tabulation of the dimensions of true Megalithic circles shows a preference for diameters integral in Megalithic yards. Each peak represents a diameter favored at several sites. (Griffith Observatory, after Alexander Thom)

Many circles were set out with diameters of 4, 8, 12, 16, 32, et cetera, MY. Each corresponding circumference was also, therefore, approximately integral. We do not know why the circumference rule was introduced, but it undoubtedly existed. The Megalith Builders attempted to adhere to it closely. The true ratio between circumference and diameter is an irrational number, 3.14159. . . . The number π does not have an exact value but can only be approximated to the desired precision. The diameters of many of the Megalithic circles had to be slightly adjusted by the builders, for the task they had set for themselves—to make diameters and circumferences integral—was impossible. We find evidence that when the Megalith Builders used a circle that did not conform well to the circumference rule, they allowed themselves to adjust the diameter slightly so that circumference became more nearly correct.

MEGALITHIC GEOMETRY

Not all Megalithic rings are true circles. One of the first variations we discovered was the flattened circle. Two major types of flattened circle are known. Both follow the same general construction, i.e., from four circular arcs, but the location of some of the centers of curvature differs from Type A to Type B. Dinnever Hill, in Cornwall, and Castle Rigg, in Cumberland, are examples of Type A, whereas Type B is represented by Bar Brook, in Derbyshire, and Long Meg and Her Daughters, in Cumberland.

Approximately 67 per cent of the stone rings are true circles. Flattened circles account for about 17 percent. Only a few of the known flattened circles do not conform to one or the other of the two types, and in some of these, like Rough Tor, it appears that a related set of rules was followed. Earlier sketches and surveys of sites that are now completely destroyed also include obviously flattened circles. Originally there must have been many more of them than remain today, and many of those that remain have not yet been surveyed.

The builders of the flattened circles did follow certain rules of geometric construction, but, curiously enough, they allowed themselves the freedom to measure out the dimensions of these shapes in non-integral lengths. The truly circular designs, by contrast, apparently were restricted to dimensions that were whole multiples of the Megalithic yard and rod. Another type of non-circular ring was constructed by these Neolithic architects, and it also had to obey the whole-unit rule. This type of ring was oval, or egg-shaped, and at least two variants, Type I and Type II, are known. The sequence of post-hole rings of Woodhenge, close to Stonehenge, is a spectacular set of Type I rings. The

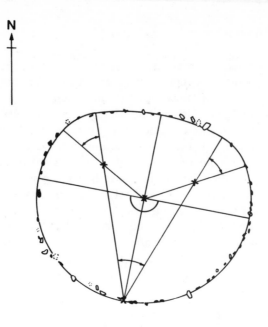

Dinnever Hill

Two major types of flattened circles are known. The circle on Dinnever Hill, in Cornwall, is a Type A. It is formed by four arcs which are all segments of circles of differing radii. (Griffith Observatory, after Alexander Thom)

Four segments of circles are also needed to draw a Type-B flattened circle, as at Bar Brook, in Derbyshire, about thirty-two miles east of the Jodrell Bank radio telescopes. (Griffith Observatory, after Alexander Thom)

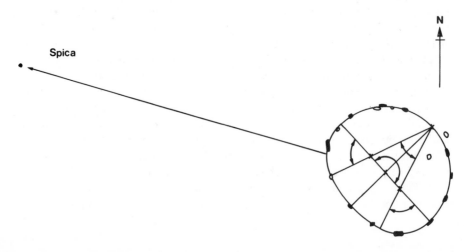

Bar Brook

Long Meg and Her Daughters, in Cumberland, is another fine example of a large Type-B flattened circle. Long Meg herself is an outlying menhir which indicates the winter solstice sunset. (E. C. Krupp)

Type II rings are not so numerous as Type I; together the two types of eggs constitute 5 per cent of all known rings.

Both types of eggs are based on right triangles. The sides of these triangles all had to be integers in Megalithic yards, and the triangles determined where the centers of curvature were to be placed. Arcs drawn from these points produce the egg-shaped rings. The best known Pythagorean triangle with integral sides is the 3–4–5 triangle. This triangle was used in antiquity to set out a right angle, as indeed it still is. Many of the Megalithic eggs were based upon the 3–4–5 triangle, but it appears in some cases other triangles were also used. Once the basic triangle was chosen, only one other dimension was needed, one of the radii. If the first radius was required to be integral in Megalithic yards, the rest would obey the same rule.

WoodHenge

The concentric post-hole rings of Woodhenge, in Wiltshire, just two miles northeast of Stonehenge, are a nest of Type I eggs whose perimeters measure 40, 60, 80, 100, 140, and 160 Megalithic yards. All of the arcs at the large end have a common center, as do the arcs at the small end. The two centers occupy the ends of the shared side of two Pythagorean triangles. The third vertex of each triangle is the center of curvature for the opposite connecting arc. (Griffith Observatory, after Alexander Thom)

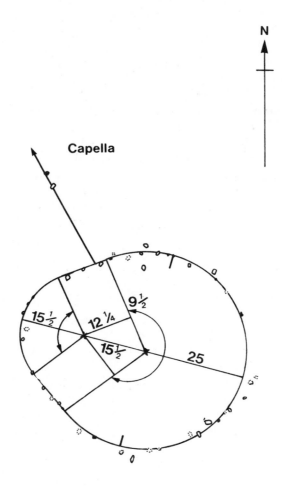

Borrowston Rig

Borrowston Rig, in Midlothian, southeastern Scotland, is a Type II egg. Here the Pythagorean triangles on which the construction is based share the same hypotenuse. (Griffith Observatory, after Alexander Thom)

At Woodhenge the basic triangle was the perfect 12–35–37 Pythagorean triangle, set out as 6–17½–18½. The builders discarded the rule that all radii must be integers but stuck to the rule that the perimeters must be multiples of the Megalithic rod. The resulting perimeters advance by 8 MR, with one step omitted. This fit was made to the positions of concrete pillars which were placed in the holes in the chalk when they were uncovered, but the pillars need not necessarily be in exactly the original post locations. This elaborate construction was probably a complicated exercise to find a ring for which π, as it were, would equal 3.0 exactly. The fifth ring from the center has a perimeter which, divided by its major axis, is equal to 3 to within 1 part in 5,000.

Why were the Megalith Builders so interested in integrals? Megalithic people did not have drawing boards and drawing paper. They did not have uniformly divided rulers nor sets of adjustable compasses. These are tools which we take for granted today. It is difficult to imagine the Megalith Builders designing and constructing their monuments without these tools, and yet that is exactly what they did.

For a drawing board the prehistoric Europeans probably used a flat stone surface, ground and polished. Some of these polishers have been found. A circle was probably drawn with trammels, that is, fixed-beam compasses. We might imagine these having been made by attaching two flint points to a stick. A logical set might have the points fixed at two, three, four, or five units apart. Many of the Scottish petroglyphs show that the smaller designs were measured or spaced in a unit of 0.816 inches. This dimension is exactly one fortieth of a Megalithic yard. All of the designs were arranged so that a set of beam compasses, without fractional lengths (excepting, perhaps, halves or quarters), could have been used to draw them.

There are two reasons for believing that the petroglyphs were drawn by the same people who set out the large stone rings. First, as already mentioned, the smallest unit, a "Megalithic inch," is exactly forty times smaller than the Megalithic yard. Second, the rings are associated with the standing stones, on which the petroglyphs are often found. The Megalith Builders had no graduated meter stick and probably used poles of fixed length to measure out their rings and large lengths. If this was so, it would have been convenient to work in integers.

Yet another carefully designed geometric shape is found among the stone rings and the petroglyphs: the ellipse. An ellipse looks like a symetrically flattened circle. Let its long dimension equal $2a$ and its short dimension equal $2b$. An ellipse can be set out with a rope by tying its ends to two fixed stakes. The stakes mark the points of the ellipse called the "foci," and if we call the distance between the two foci $2c$ the length of our rope is $2(a+c)$. A third stake is used to inscribe the

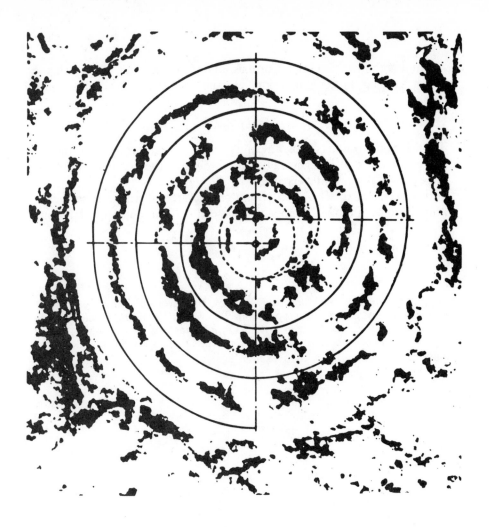

Ellipse halves of increasing size are paired on this petroglyph at Knock, in Wigtownshire, Scotland, to form a spiral pattern spaced in intervals of 1 Megalithic inch. (R. W. B. Morris and Alexander Thom)

elliptical shape on the ground. The staked loop of rope is pulled tight as the third stake is slid along the ground. Ellipses of many different shapes are possible, and the flatter the ellipse, the greater its eccentricity is said to be. If the eccentricity is reduced to o, the ellipse becomes a circle. The eccentricity is simply c/a. When the ellipse becomes a circle, the two foci coincide, and the distance between them equals o.

The long dimension, or major axis, of the ellipse is 2*a*. If we call the minor axis, or shorter dimension, 2*b*, the following relationship holds true.

$$a^2 = b^2 + c^2$$

The Megalith Builders took advantage of this relationship and tried to make *a*, *b*, and *c* all integers. They tried to make the perimeters of their large elliptical rings integers also.

It is very easy to set out an ellipse on the ground with a rope, but it must have been almost impossible to use this technique on a flat stone

One of the many stone rings on Dartmoor, in Devonshire, that is a perfect ellipse, the Postbridge ring, or Soussons Plantation Retaining Circle, as it is also called, has a major axis of 10½ MY, a minor axis of 10.06 MY, and a distance between foci of 3 MY. (Griffith Observatory, after Alexander Thom)

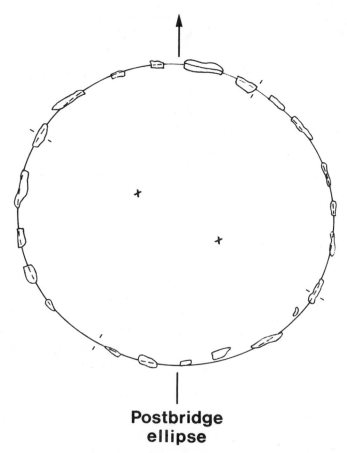

**Postbridge
ellipse**

surface. Pins cannot be stuck into stone to mark the foci. A complete set of beam compasses, advancing, perhaps, in quarter units, could enable the designer to set out enough points geometrically to draw the ellipse, however. The rest of the ellipse could have been filled in by eye, and probably this is what was done. In the field the builders could have set out an ellipse of any size, but on the small scale, on stone, their dimensions must have been integers.

Megalithic geometers evidently preferred to use integral lengths, and it is possible that this preference was motivated by a desire to avoid graduated measuring sticks. Straight lines, such as diameters and radii, were made integral in Megalithic yards, but perimeters and most very long distances were measured, almost universally, in integral Megalithic rods. Probably the attraction of the ellipse to these people is its additional variable, the distance between foci. It would be easier to obtain a circumference with a desired value by adjusting a, b, and c. The ideal Megalithic ellipse would have $2a$, $2b$, and $2c$ integral in Megalithic yards and the perimeter integral in Megalithic rods. This ideal is mathematically unobtainable, but when we consider the resources available to these people, it is surprising how often they found good approximations.

In true circles, flattened circles, egg-shaped rings, ellipses, and other compound designs, as in Avebury, the Megalith Builders experimented with geometry and imposed rules of measurement. We do not know how these ideas related to their other institutions, but for some reason the mathematical principles they explored were sufficiently important to them to be committed to stone.

MEGALITHIC ASTRONOMY

We do not know what conceptions of the cosmos were held by the Megalith Builders. Nor do we know if they understood the relation between the earth, the moon, and the sun. We do know, however, that they developed a method of observation which could have enabled them to go very far to extract even the subtlest patterns from constant movements of the sun and moon.

The Megalith Builders used rows of standing stones, or "menhirs," and large, single menhirs viewed from smaller stone backsights to indicate significant rising and setting points of the sun, moon, and certain stars. Alternatively, a notch on a ridge located miles from the observation post was used as a foresight. The islands of Boreray and Cara in Scotland's Hebrides, the cliffs of Hoy in the Orkneys, and Bass Rock off the Firth of Forth in the North Sea were used as foresights. Flat-

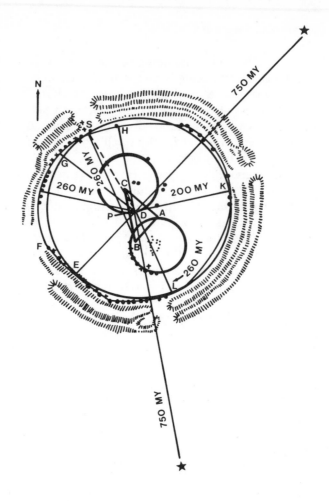

The construction of the great ring at Avebury, in Wiltshire, may have been based upon a Pythagorean triangle and with radii set out in Megalithic yards. These radii generated six arcs which were drawn to achieve, perhaps, a total perimeter of 1,300 MY, or 52 MR. The inner circles at Avebury have diameters of 125 MY and perimeters of. 157 MR and are identical in size to the Ring of Brogar, far to the north, in Orkney. (Griffith Observatory, after Alexander Thom)

sided stones often pointed to a notch, as at Ballinaby on Islay in the Hebrides. The centers of some circles were used as backsights for an outlying foresight stone. Hundreds of such lines have been measured. The exact compass point, or azimuth, and height to the true horizon profile, or altitude, have been carefully surveyed. From these data astronomical positions above and below the celestial equator—that is,

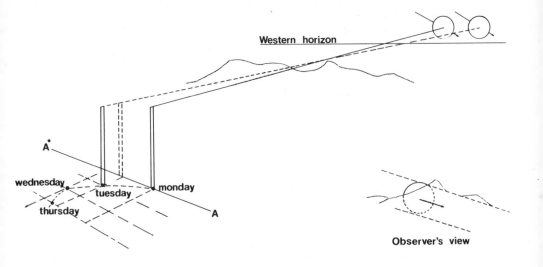

Western horizon

A*

wednesday
tuesday
monday
thursday

A

Observer's view

Method of Observing

*Without telescopes or calibrated measuring instruments, prehistoric astron-
omers had to use the horizon to determine the progress of astronomical cycles. Two things were needed: a place to stand and a place to look. For example, the astronomer marked with a stake the necessary place to stand to
make the sun appear to set behind a distant natural foresight on the horizon. Each day, as the summer solstice nears, the sun appears to set to the
right of the foresight, and each day in turn the astronomer moves the stake
to the left to achieve alignment with the mountain peak and the setting
sun. The leftmost stake position marks the time and place of the actual solstice alignment. (Griffith Observatory, after Alexander Thom)*

declinations—were calculated. A study of these declinations, presented
in graphic form, shows which astronomical phenomena interested people in the Megalithic era.

Just as before, when each measured diameter and circumference of a
ring was represented by a small area on a chart of all possible lengths,
each astronomical declination is represented by a small Gaussian area,
whose shape indicates the reliability of the determination. (A Gaussian
area is shaped like a small hill and its exact form follows from the laws
of random errors.) A well-determined line is tall and narrow. Less well-
determined alignments are wide and low. We find the declinations

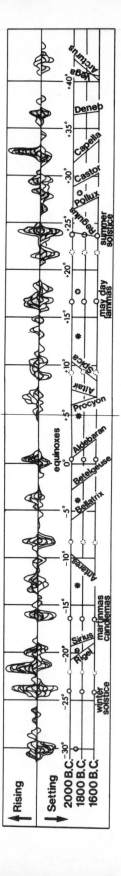

Megalithic alignments preferentially coincide with astronomical events, primarily sunrises, sunsets, moonrises, and moonsets. Declination, or angle above or below the celestial equator, is plotted horizontally, and peaks build up at significant astronomical declinations of the sun and the moon. Some stellar alignments may also be indicated by this statistical diagram. Martinmas and Candlemas share the same declination, and May Day and Lammas share another. Both the vernal and autumnal equinoxes also occupy the same declination. (Griffith Observatory, after Alexander Thom)

from observed alignments piling up significantly, and it is obvious, even without statistical analysis, that the best lines (shaded areas, Class A, or well indicated, alignments) form definite groups that cannot be explained by chance. These lines show peaks in this way only when declinations are plotted. Therefore, the majority of alignments must be astronomical. The most impressive lines to emerge from such a study are solar alignments for certain calendar dates and lunar alignments at the standstills. In addition, some stellar alignments on first magnitude stars may also be present for the epoch of construction of the Megalithic monuments. It is more difficult to confirm stellar orientations, for they depend strongly upon the date of construction, and dates for most of the megaliths are not that well known.

Further confirmation of an astronomical interpretation of some Megalithic alignments is provided by a detailed look at the sun's behavior. In Megalithic times the sun's declination at the solstices was about ±23.91 degrees. The declination of the upper limb of the rising sun, when it first appeared on a *level* horizon, would have been about 0.22 degrees greater than this, however, and the lower limb's declination would have been about 0.22 degrees less. We would expect the Gaussian areas to pile up to a maximum toward the one edge of the sun's disc, which is drawn in on the declination scale for comparison, if the actual limbs were observed. This appears to have been the case at the solstices, and common sense, of course, would have dictated it.

Certainly a diagram of possible astronomical alignments inevitably must carry a number of spurious lines. Accidental, intrusive lines cannot explain, however, the concentration of rising Gaussian areas that peak at +32.5 degrees. These lines must belong to the bright star Capella, as observed in roughly 1800 B.C. Several other first magnitude stars also appear to have several alignments for a date between 1800 and 2000 B.C. These stars may well have been used as clocks, or timekeepers. Perhaps a night watchman was warned by a star's rising or setting to wake the astronomer early enough to observe the rising of the sun or position of the moon.

THE MEGALITHIC CALENDAR

It is easy to know that the solstices are approaching, for the rising sun moves closer and closer in its extreme position on the horizon. It is very difficult, however, to pinpoint usually the exact day of the solstice. The sun changes its position on the horizon only slightly during the days before and after the solstice. It would take a highly precise instrument to peg the exact day of solstice.

Suppose we have found a place to stand from which the sun appears to set over a well-defined mountain peak in the northwest on the date of the summer solstice. If the peak is a considerable distance away, even a small movement from the proper position will place us so as to miss the event on the proper day. If the backsight and foresight are chosen carefully enough, it is therefore possible to come very close to an observation of the exact day of solstice. The system could be refined even further by making observations of the setting sun several days before and after the yet-to-be-determined solstice. The observer could place a stake in the ground each day on a line perpendicular to the line of sight to the horizon. After the set of observations was completed, the observer would know that the stake farthest to the left, as the setting horizon was faced, approximated most closely the day of the summer solstice. It would even be possible to refine this procedure even more by geometrically projecting where the actual leftmost position must be, even if the time of direct observation prohibited seeing it. Indeed, it is unlikely that the exact hour of the solstice would coincide with the sunset, but observations of this type would permit the Megalith Builders to approximate it and to locate the correct position for a marker had sunset and solstice occurred together.

There are many solar sites of sufficient precision to permit exact determination of the solstice. Ballochroy is a simple alignment of three stones on the Kintyre peninsula in Scotland. Peaks to the northwest, on Jura, and to the southwest, on the small island of Cara, provide good solstitial lines.

Kintraw is about forty miles north of Ballochroy. A 12-foot menhir and the ruins of a large cairn and small ring remain. The line from Kintraw to Jura indicates a notch in the horizon profile, twenty-eight miles away, which marks the winter solstice sunset. We noticed, however, that the notch could not be seen from the cairn, for a foreground ridge blocks the view. Looking backward along the alignment we discovered a small flat ledge on the steep hillside across the gorge. A large boulder was found on the ledge and on the solstitial alignment. The notch can be seen from this ledge, which provided sufficient room in the right direction (across the line of sight) to permit observation several days before and after the solstice. E. W. MacKie, a Scottish archaeologist, has excavated this platform and shown that it was formed by artificial means. It has no other obvious function than as a place from which solstitial observations are to be made.

Boreray, the most northerly island of the St. Kilda archipelago, is far out in the Atlantic from the Outer Hebrides. It was used as a foresight from four menhirs, each many miles away from one another. These menhirs, on the west coasts of several Hebridean isles, in conjunction

The right end of the island of Cara, west of Kintyre, as seen looking south-west down the Ballochroy alignment, precisely marks the winter solstice sunset. (E. C. Krupp)

with Boreray, provide several solar alignments, including a summer solstice sunset. Boreray is itself very precipitous and 1,245 feet high. Its peak, seen even from forty to sixty miles away at the menhirs, projects above the sea horizon. Here also some kind of interpolation procedure must have been used. The movement of the sun along the horizon in a single day at the equinox is greater than the diameter of the orb. At one sunset the sun might be south of the peak and by the next may have moved past it to the north. Only one of the menhirs suggests a solstitial alignment, however. The others appear to be related to other dates and may well have been part of the Megalithic calendar.

The solar, or tropical, year is defined and divided by the solstices, winter and summer. We find it convenient to subdivide the year still further, however, and doubtless the Megalith Builders did too. Two other astronomical phenomena correspond to convenient dates to quar-

BALLOCHROY

The central stone of Ballochroy is shown on the right in this drawing. It may indicate the northernmost pap of the island of Jura, nineteen miles away, and the summer solstice sunset. (Griffith Observatory, after Euan W. MacKie)

From Kintraw the island of Jura provides a highly precise foresight for the winter solstice sunset. The stone at the right is the tall menhir remaining at Kintraw, and the peaks of Jura form the distant horizon profile. The pole at the left marks the location of a prehistoric socket discovered in the center of the ruined cairn. (Griffith Observatory, after Euan W. MacKie)

KINTRAW

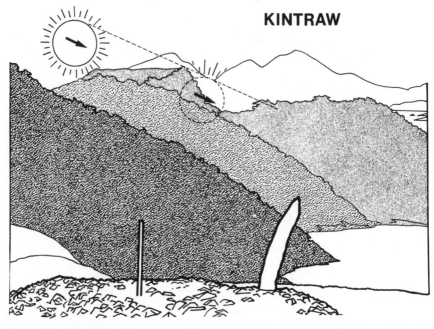

A narrow flat ledge on the steep hillside at Kintraw is perfectly positioned for astronomical observation and appears to have been constructed artificially. (E. C. Krupp)

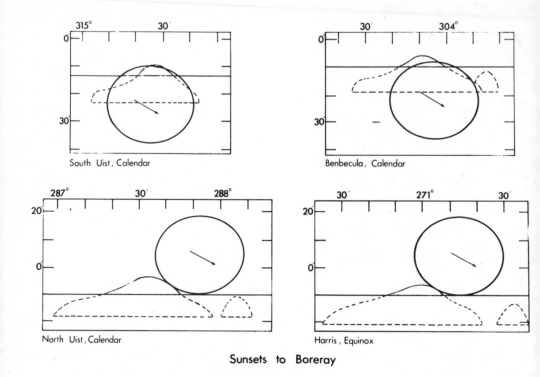

South Uist, Calendar

Benbecula, Calendar

North Uist, Calendar

Harris, Equinox

Sunsets to Boreray

The island of Boreray, in the Hebrides, acts as a foresight for several solar alignments as viewed from four different menhirs on other islands in the Hebrides. (Griffith Observatory, after Alexander Thom)

The Megalith Builders quite sensibly calibrated their equinox alignments on the sun's position on the days that bisected the year into intervals of an equal number of days. The true equinox occurs when the sun actually crosses the celestial equator, but the Megalithic equinox alignments are offset in the correct direction and by the correct amount for calendar bisection. (Griffith Observatory, after Alexander Thom)

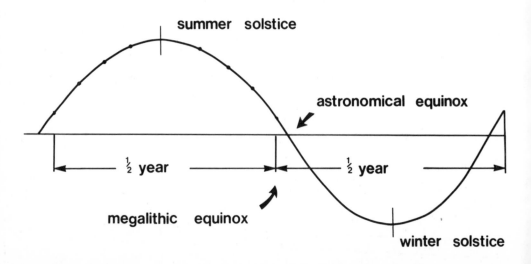

summer solstice

astronomical equinox

½ year

½ year

megalithic equinox

winter solstice

ter the year: the vernal and autumnal equinoxes. The moment of equinox occurs when the sun crosses the celestial equator, but this is not a very practical definition for the observationally oriented Megalithic peoples. It is far more likely that the year would have been further subdivided to put an equal number of days into each quarter. When the equinoxes are defined in this way, they do not correspond to the declination of o degrees, where the sun crosses the equator, but to a declination of ½ degree.

Were the earth's orbit about the sun truly circular, the true equinoxes would quarter the year into equal intervals. Instead, the earth's orbit is elliptical. Therefore, the time from spring equinox to fall equinox is slightly longer than the time from fall back to spring. If the Megalith Builders established equinoxes by day counts rather than by declination, we should expect equinoctial alignments to be systematically too high by about ½ degree. This seems to be verified.

Still further subdivision of the year might be desired. Eighths of a year could be obtained by choosing four more dates: one in summer, one in fall, one in winter, and one in spring. The time interval between these dates must be exactly a quarter of a year. These four dates, in coordination with the two solstices and the two Megalithic equinoxes, give an eight-"month" year. The additional four calendar dates coincide, quite interestingly, with traditional Old English and Scottish holidays: May Day (or Beltane), Lammas, Martinmas, and Candlemas. These dates were also used in pagan times by the Celtic peoples of Britain, and, of course, their antiquity may be even greater. Candlemas (February 2) has been replaced by Groundhog Day in the United States.

Our diagram of astronomical alignments shows concentrations of the Gaussian areas into peaks at the correct solar declinations for the four midcalendar dates. There appear to be peaks at yet four more positions, which would divide the year into sixteen "months." A careful analysis of the declinations reveals a possible calendar with eleven intervals of twenty-three days, four of twenty-two days, and one of twenty-four days. This has every appearance of a successful solution to the very difficult task of dividing the 365 days of the solar year into a convenient number of very nearly equal smaller units of time. Certain otherwise unexplained alignments support yet another subdivision, with thirty-two intervals of eleven or twelve days, but further examination and more data will be required to determine if it is real.

Without a calendar it is impossible to predict any future event, be it football game or solstice. The Megalithic calendar permitted these Neolithic and Bronze Age people to make astronomical predictions accurately. It is even likely that the Megalith Builders knew of the neces-

The rows of small stones at Learable Hill, on the border of Sutherland and Caithness, in Scotland, align with the Megalithic equinox. (E. C. Krupp)

sity to intercalate a leap year day every four years. Their knowledge of
the sun's movements would have been sufficient to make them aware of
this. In fact, if they did use a solar calendar as described above, the
need for leap year would have been forced upon them after five to ten
years of their calendar's use.

It is quite possible, through the use of a distant foresight and markers
on several evenings or mornings of observation, to make very precise
astronomical observations. Had the Megalithic astronomers taken a
step back each sunset, as they placed their stake in the ground, the sequence of stake positions would have generated a curve. Just as the
markers on the line across the line of sight reach an extreme position
for the solstice, a curve traced by stepping a pace or two backward, as
described, would come up to a maximum and decrease again with the
changing declination of the sun. The accuracy available to Megalithic
astronomers through this simple method would have far exceeded the
accuracy of any known measuring instrument of antiquity. No comparable accuracy using an instrument was possible until the invention of
the telescope. There was, however, an important snag—atmospheric
refraction. Refraction increases as the altitude of any celestial object
falls until, on the horizon, refraction can amount to ½ degree. For this
reason, when we see the setting sun, we are really looking at the image
of its disc that has been displaced above the horizon by the air around
us. Really precise measurements of position can be affected by refraction, which unfortunately can vary from night to night—with temperature, atmospheric pressure, cloud cover, and many other details of the
atmosphere. Interpretation of many Megalithic sites becomes very
difficult when we try to attain an accuracy of less than a minute of arc.

MEGALITHS AND THE MOON

The moon could have been as likely a target for the Megalith Builders
as the sun. Observation could have been carried out in the same fashion
as for solar phenomena. The rapidly moving and quickly changing
moon would be observed on the horizon. The upper limb, the lower
limb, or the center of the disc might each have been used to mark the
progress of its motions. An observer would again mark, by placing a
stake in the ground, the correct bearing of the ray from the moon's
lower limb when it touched the foresight. An assistant would mark the
bearing of the upper limb, when it grazed the foresight, with a second
stake. A point on the ground midway between the two stakes would
record the bearing of the midorb. Such observations require fairly level
ground and considerable patience. The sun's diameter is almost con-

stant, at 32 arc minutes, but the moon's varies from 29.4 arc minutes to 33.4 arc minutes in a month. Over a month the average lunar diameter is 31.03 arc minutes. The moon is much closer to the earth than the sun, and variations in the moon's distance show up more dramatically.

Looking once again at the chart of observed astronomical alignments, we see that many lunar lines are also present. These mark moonrises and moonsets for the extreme positions of the moon at the time of the lunar standstills. The declination of a limb of the moon will be given by a combination of the following quantities:

ϵ, the obliquity of the ecliptic;
i, the inclination of the moon's orbit;
Δ, the inclination perturbation; and
s, the moon's semidiameter.

By deducing average values of these numbers which best suit the most reliable lunar lines and by correcting for nighttime refraction and for parallax (a shift in apparent position caused by the moon's varying distance from earth), it is possible to reconstruct how careful lunar observations may have been made. A mathematical comparison between the observed lines and the theoretical position for lunar alignment gives the following results:

$\epsilon = 23$ degrees 53 minutes
$i = 5$ degrees 9 minutes
$\Delta = 9$ minutes
$s = 16$ minutes.

The values i, Δ, and s agree with their modern values, as they should. The obliquity of the ecliptic, ϵ, is slowly decreasing at a known rate. Today the angle between the celestial equator and the ecliptic is 23 degrees 27 minutes. The value obtained for ϵ from the megaliths indicates construction and use at an earlier date, but it is not determined accurately enough to assign a definite date. In 1700 B.C. ϵ was 23 degrees 53 minutes, and this is sufficiently close to the time archaeologists estimate the megaliths were erected for consistency with the calculated astronomical date. This astronomical determination can be wrong by several centuries, however, because it is affected by parallax in an ambiguous manner.

About forty good lunar alignments are established. That given by the menhir at Ballinaby is typical. Its large height suggests a lunar site, and its long sides are oriented on a notch which indicates the major standstill moonset and permits observation of the effect of the small inclination perturbation, Δ.

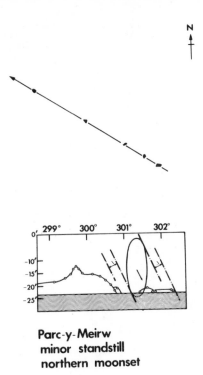

**Parc-y-Meirw
minor standstill
northern moonset**

The alignment at Parc-y-Meirw, in Wales, coincides with the minor stand-still northern moonset. (Griffith Observatory, after Alexander Thom)

Parc-y-Meirw in Dyfed (formerly Pembrokeshire), Wales, is a note-worthy lunar backsight because the distance to the foresight is so great, ninety-one miles. The foresight is not often visible because of poor weather in our modern era. The site consists of four large menhirs and is probably lunar. The ground rises and prohibits observation to the southeast. The minor standstill moonset could have been observed, however, to the northwest. Mount Leinster in Wexford, Ireland, and some surrounding ground is always above the sea horizon and is in the correct lunar orientation. Clear skies would have been required to see this foresight silhouetted against the moon. The climate in Britain must have been better in Megalithic times than today. One interested observer indicated that he had seen the Irish hills from Parc-y-Meirw on a clear day. Its use as a lunar observatory is therefore quite possible.

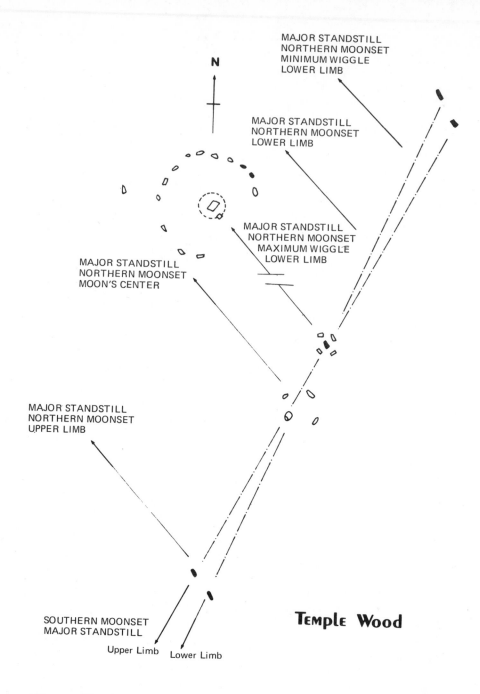

The peculiar elongated arrangement at Temple Wood, in Argyll, Scotland, could have been used to make precise observations of northern and southern moonsets at the major standstill. (Griffith Observatory, after Alexander Thom)

The notch behind the southwestern pair of menhirs at Temple Wood could allow observation of the inclination perturbation. (E. C. Krupp)

Near Kilmartin in Argyll, Scotland, stands an impressive line of menhirs near Temple Wood. Two moonset notches, one for the northern major standstill declination and the other for the southern major standstill declination are present. The menhirs are in an X-shaped arrangement and mark points from which upper and lower limbs and disc center would have been observed. This observatory also permits detection of the inclination perturbation.

A most impressive circle, the Ring of Brogar, is found on Mainland, in the Orkneys. The diameter of this ring is 340 feet, or 125 MY. This is the same size as each of the two inner circles of Avebury, far to the south in England. The circumference of the Ring of Brogar is 392.5 MY, or 157 MR.

Brogar, at first sight, does not appear to have much astronomical significance. A system of small earth cairns outside the ring and an oriented outlier, the Comet Stone, suggest, however, there is more here than meets the eye. A careful survey of the horizon profiles revealed four natural foresights, and these make Brogar a remarkable lunar observatory. The minor standstill southern moonrise and moonset and the major standstill northern moonrise appear to have been used for observation. A major standstill northern moonset foresight has just recently been tentatively identified. The various cairns and mounds are the backsights, and together they give nearly a dozen declinations of the moon. Recent data we have acquired regarding refraction and regarding the moon's movements in the second millenium B.C., provided by Dr. A. T. Sinclair of the Royal Greenwich Observatory, should eventually permit extraction of the actual observing procedure from the remaining features of Brogar.

MEGALITHIC ECLIPSE PREDICTION

It is obvious that the Megalith Builders knew of the 18.61 year lunar standstill cycle and measured it, but this activity can hardly justify the high precision of the Megalithic lunar observatories. Their astronomical potential was far greater, great enough to permit prediction of eclipses. As it happens, an eclipse can only occur when the sun and the moon are sufficiently close to a node of the moon's orbit.

Relative to a lunar node the sun circles the celestial sphere in 346.6 days, an eclipse, or draconic, year. The critical periods, during which an eclipse can occur, repeat every 173.3 days, or half an eclipse year.

The moon responds to the gravitational force of the sun, by a small perturbation of the tilt of the lunar orbit. This is just one small effect of the sun on the moon's total behavior. This variation, first 9 arc min-

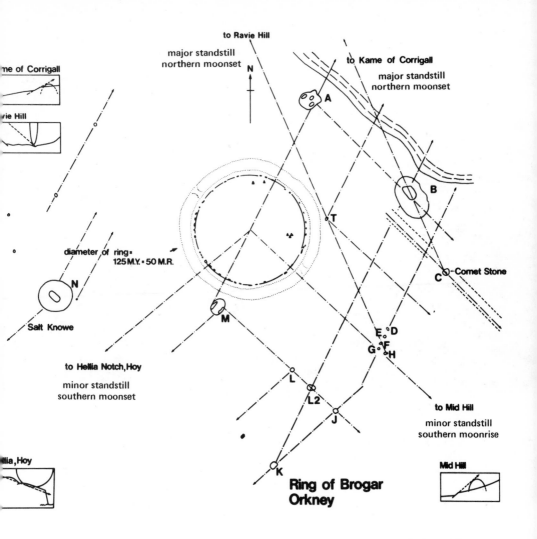

**Ring of Brogar
Orkney**

Astronomy at Brogar, on Mainland, Orkney, involves the mounds around the stone circle and not the stones themselves. The basic alignments are major standstill northern moonrise, minor standstill southern moonrise, and minor standstill southern moonset, with several adjustments for the moon's semidiameter and the inclination perturbation. (Griffith Observatory, after Alexander Thom)

The line from Brogar's Mound A to Mound B and southeast to Mid Hill indicates the minor standstill southern moonrise for the moon's upper limb. (E. C. Krupp)

utes up, then 9 arc minutes down, and then 9 arc minutes back up, completes its cycle during the time the sun moves from one node to the opposite node. This interval is, of course, one half of an eclipse year. The maximum tilt always occurs when the sun is passing either of the nodes, and the little oscillation in the moon's motion, as noted in Chapter 1, is the inclination perturbation.

During the course of the year the sun's declination oscillates between two limits, 23½ degrees above the celestial equator and 23½ degrees below. The moon is inclined about 5 degrees to the ecliptic, and its maximum declination in a given month varies from ±28½ degrees at the major standstill to ±18½ degrees at the minor standstill, 9.3 years later. The entire standstill cycle takes 18.61 years.

Even at the standstills the moon in no sense stands still, but for about a year the limiting declinations do not vary by more than 20 arc minutes, or by about two thirds of the moon's apparent diameter. The last major standstill took place in 1969. The effect of the inclination perturbation, in the form of a small oscillation on top of the moon's average declination extrema, is particularly visible during the time of a standstill. In fact, the only time the Megalith Builders could have observed it was as an added shift of the moon's position on the horizon on top of its standstill extreme. It would have been worth their while to detect this subtle shift, however, for eclipses can occur only at the time when the perturbation is at its maximum. As already noted, the date of this maximum can be determined by observing the moon's shifting position at moonrise and moonset at the standstills. The Grand Menhir Brisé, a 340-ton standing stone that once probably stood 60 feet in the air near Carnac on the coast of Brittany, France, may have been used as a universal lunar foresight from several, if not all eight, standstill backsights. Once the occurrence of maximum perturbation was observed, it would have been a simple matter, with the help of the Megalithic calendar, to count the days until the next eclipse season, provided the period of inclination perturbation was known. Eclipses can only happen when the oscillation is contributing its greatest effect to the current limits of the moon's positions on the horizon. Between standstills the inclination perturbation is still there, but it is lost amid the other movements of the moon. Only at the standstills could it actually have been observed by the Megalith Builders. From then until the next standstill the calendar had to carry the responsibility for accurate prediction of eclipse seasons.

The moon's declination changes so fast, from nearly +29 degrees to −29 degrees in half a month, that during any particular lunar month, or lunation, the moon may come up to its maximum and retreat again in the time between two successive observations, which by necessity are separated always by about a day. It is possible to deduce when the actual moment of maximum displacement occurs and the location of the correct stake position for this moment by a process of extrapolation. A careful and clever treatment of the observed positions of stakes marking the moon's position relative to the horizon on two or three days around a maximum declination, be it the monthly limit, the standstill limit, or the inclination perturbation limit, could have allowed the Megalith Builders to calculate the desired time and stake position of the actual maximum. Ancillary equipment would be needed to perform this calculation geometrically, but the necessary features appear to be present at many of the lunar sites.

One technique for determination of the moon's precise extreme involves the use of similar triangles and could have been used at Temple Wood. Any extrapolation method depends upon the distance the stake

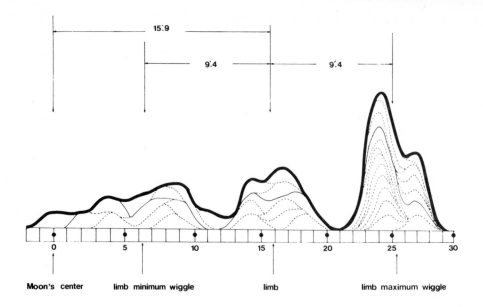

Difference between Observed Declinations and $\pm (\epsilon \pm i)$

A *comparison of the directions of Megalithic lunar alignments with the actual celestial positions of the moon's center at major and minor standstills is
convincing evidence that precise lunar alignments were intended. The
difference in declination between a measured alignment and the moon's actual position is plotted horizontally. Certain preferred angular differences
show up as peaks which correspond in angular distance from the moon's
center (o) to its upper and lower limbs at maximum, minimum, and zero
inclination perturbation. These results suggest fine adjustment in the alignments to pick up the effect of the perturbation.* (Griffith Observatory, after
Alexander Thom)

moves from day to day, and this, in turn, depends on the distance to
the horizon. The extreme stake position can be found with the help of
certain standard distances, and just the distances that are required for
both notches at Temple Wood separate certain features of the monument, as though those features were a permanent record of the needed
distances.

A second possible method for pinning down the moon's extreme position involves sectors of circles. Many fanlike arrangements of stones,

A MAJOR STANDSTILL

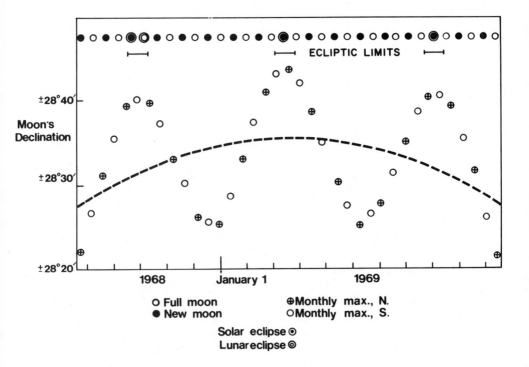

In this diagram the moon's actual monthly declination maxima, or extremes, to the north and south of the celestial equator, are shown for the
1968–69 major standstill as open (monthly maximum north) and crossed
(monthly maximum south) circles. The drawing is set up in a way that permits marking positive and negative declinations on the same vertical scale.
If there were no inclination perturbation, all of these circles would fall upon
the dashed arc, which represents changing declination of an unperturbed
moon at the standstill. Instead, the circles fall upon a wave-shaped path that
oscillates above and below the dashed arc. The moon, at the monthly extremes, is carried above the dashed arc during one half of the 173-day perturbation cycle, and it is carried below the dashed arc at the monthly extremes during the cycle's second half. The dates of new and full moons and
of eclipses are indicated by coded circles at the top of the diagram. It can be
seen that eclipses occur only when the moon is most perturbed, that is, at a
crest of the wave. (Griffith Observatory, after Alexander Thom)

which meet, in shape and size, the extrapolation requirements, are found at several lunar sites in Caithness, in the north of Scotland. Similar, although more complex, rows of stones dominate the landscape at Carnac. No other satisfactory explanation of the stone rows and fans has been suggested. The geometric extrapolation method that is mentioned here corresponds exactly to what is found in these curious sites, which seem always to be present beside or near obvious lunar backsights.

The stone rows at Carnac are of particular interest, for here in Brittany are the remains of two of the most elaborate and highly developed Megalithic lunar observatories known. For the first of these, the massive menhir, 16 feet high, of Le Manio stood as the universal foresight. It was viewed from seven or eight marked lunar and solar backsights. The nearby rows of Le Ménec extend for over 0.6 mile and consist of 1,600 stones. Originally, a Type I egg fitted onto the west end of the alignments, and sparser remains of what may have been a Type II egg are on the east end. Stones in each of the original twelve rows were spaced 1 MR apart. The rows of Kermario, just northeast of Le Ménec have a more complex design but were also spaced in Megalithic rods. While the use of the Kermario geometry is still unknown, it appears that the Ménec alignments were used as stone graph paper to extrapolate the position of the moon from observations over the largest megalith in Europe, the Grand Menhir Brisé.

The Grand Menhir Brisé, as previously noted, served as the universal foresight for several of the possible standstill backsights. At present five actual backsights have been located. Although the Grand Menhir is now fallen and broken, probably by earthquake, it must once have been visible to prehistoric astronomers stationed all around the Quiberon Bay. In conjunction with the usual type of stone fan, at St.-Pierre and Petit-Ménec, for example, and with the complicated geometry of the Le Ménec alignments, the Grand Menhir would have given the Megalithic astronomers a means to solve the extrapolation problem with three clear nights of observation at one of the monthly declination extremes near the standstill.

There are many interesting Megalithic features in the Carnac landscape, and of these the rectangular cromlech at Crucuno is worth mentioning. Its short sides are aligned north–south, and a 30-by-40 MY rectangle fits the plan. The diagonal of the rectangle is therefore 50 MY, and in this structure we have the familiar 3–4–5 Pythagorean triangle. The diagonals are also astronomically significant. They indicate sunrise and sunset on the summer and winter solstices. An arrangement this concise and symmetric is only possible at latitude 47 degrees 37.5 minutes. Crucuno's latitude is 47 degrees 31 minutes. Its builders were

It is not likely the moon's extreme positions would always be occupied exactly when the moon is on the horizon, rising or setting. It would be possible to still determine the correct stake position and time for this extreme using auxiliary triangles or grids. The Hill o' Many Stanes at Mid Clyth, in Caithness, Scotland, is one of several such grids which have heretofore defied explanation. (E. C. Krupp)

only six or seven miles south of the ideal position. Unfortunately, there are no standing outliers to verify the alignments that are suggested by the relatively short dimensions of the rectangle. Similarly, it may be that stones in the sides of the rectangle indicate some of the lunar standstills' risings and settings, but the small size of the structure prohibits any categorical conclusion.

Two types of observatories seem to have been in use in Megalithic times. At Brogar, Temple Wood, Ballochroy, Kintraw, and Stonehenge the observer stood at or near the same spot and looked out to one or more distant foresights. For Boreray, the Grand Menhir Brisé, and Le Manio the plan was turned inside out. These markers worked as universal foresights for several backsights.

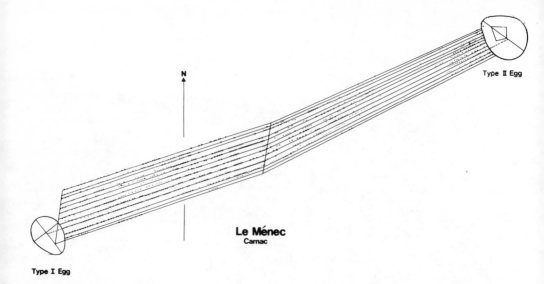

N

Type II Egg

Type I Egg

Le Ménec
Carnac

Across the English Channel, on the south coast of Brittany, the vast grid of stones of Le Ménec are but one component of the gigantic complex of megaliths at Carnac. Two familiar shapes, a Type I egg and a Type II egg, are located at either end of the rows, which are spaced from each other in Megalithic units, as are the stones in a single row. The alignments could have been used as Megalithic graph paper to extrapolate the extreme positions of the moon. (Griffith Observatory, after Alexander Thom)

At Crucuno, near Carnac, the stones are arranged as a rectangle. The diagonals may have been intended to point to the four solstice positions. The other lines, shown here, may have also been intended, for they roughly coincide with the eight lunar standstill positions. (Griffith Observatory, after Alexander Thom)

N

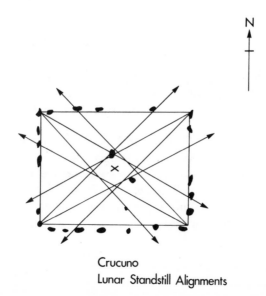

Crucuno
Lunar Standstill Alignments

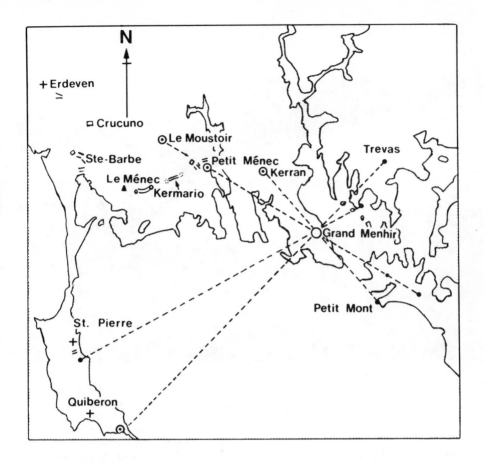

Lunar Standstills at Carnac

Several sites near Carnac, all around the Quiberon Bay, appear to be appropriately located to permit observation of the lunar standstills over a universal foresight, the Grand Menhir Brisé. (Griffith Observatory, after Alexander Thom)

The Grand Menhir Brisé, near Carnac, is a megalith of truly monumental proportions. When standing, the sixty-eight-foot menhir probably towered sixty feet above the ground. It is not known when the stone toppled, but when it did, it broke into five pieces (one is now missing). Those that remain weigh a total of 340 tons. This huge stone could have been seen from around Quiberon Bay and represents one of the most ambitious observatories in Megalithic Europe. (E. C. Krupp)

In the high precision and complexity of the Megalithic observatories we see the potential for formulation of an accurate calendar and for eclipse prediction. The full meaning of the Megalithic rings may elude us, but we can recognize a conscious sense of design and an adherence to certain rules of construction. The dimensions of Megalithic monuments imply the existence of a well-used and highly valued system of measurement. Taking all of these aspects of the prehistoric landscape into consideration, we are forced to conclude that the people who planned them, experimented with them, and used them were far more organized and sophisticated than we have previously assumed. We must now recognize these people for what they were, the originators of a unique and independent culture whose motives and dreams we have only begun to sense.

CHAPTER THREE

The Stonehenge Chronicles

E. C. KRUPP

It may come as a surprise to discover that the many prehistoric stone rings and standing stones in Britain and France are so elaborate in design and astronomical potential, but Stonehenge, the most famous Megalithic monument of all, has long been associated with the sun. Gerald S. Hawkins was responsible in the last decade for the last great wave of popular interest in Stonehenge astronomy. Hawkins reiterated a claim made as early as 1771 that the Heel Stone, as viewed from the center of the circle, marks the sunrise at the summer solstice. Hawkins found many other alignments on the sun and the moon and captured the imagination of the public by using a computer to verify his results. Even more appealing to many readers was his claim that Stonehenge was itself a computer, a predictor of eclipses.

Others have also looked at Stonehenge astronomically—Norman Lockyer, Fred Hoyle, C. A. Newham, and Alexander Thom among them. Their various interpretations are explained, compared, and evaluated in the pages which follow. Some of Hawkins' theories are treated skeptically here, but anyone at all curious about Stonehenge is indebted to Hawkins for stimulating the discussion so well and for helping to create a climate in which further studies of Stonehenge astronomy might prosper.

Whatever the true purpose of Stonehenge may have been, the old stones continue to stimulate our imaginations. Approximately 700,000 visitors soak in its mystery each year. Latter-day Druids have held their solstice ceremonies there for many summers in search of that elusive rising sun. Astronomical theories of Stonehenge seem to return periodically, like certain comets, and controversy regularly tails close behind.

The Heel Stone looms above Stonehenge when viewed from the roadside of the busy A344. (E. C. Krupp)

If you approach Stonehenge today by automobile from the northwest on the A344, you are on top of the stones and past them before you have had a chance to realize where you are. The monument is about 200 feet from the highway and seems small within the vistas across the chalk downs of the Salisbury Plain. The celebrated Heel Stone is, in fact, right next to the road.

Alternatively, approaching from Amesbury, three miles to the east on A303, Stonehenge is visible on the horizon at the crest of a rise in the Western Downs. From this vantage Stonehenge looks even smaller, but in neither case does it look out of place. It is the roads, the automobiles, the tourists, the hitchhikers, and the fences that seem not to fit.

Stonehenge is integrated into the open landscape by the intentions of those who built it. The "hanging stones" do not loom from the earth as if to dwarf the countryside. Rather they seem to mark a spot. Inside the circles one feels not so much their awesome size as the mystery of their use.

In the seventeenth century Samuel Pepys wrote in his famous diary that the stones of Stonehenge were

> as prodigious as any tales I ever heard
> tell of them, and worth going this journey
> to see. God knows what their use was . . .

Pepys was one of the few to marvel at Stonehenge without contributing his own speculation of its origin and purpose. Druids, Danes, Romans, Saxons, Phoenicians, Minoans, Egyptians, Hyperboreans, Irish, Mycenaean Greeks, Atlanteans, and ancient astronauts have all been proposed as its builders.

We now know that the antiquity of Stonehenge so far exceeds all written accounts of it that only the techniques of modern archaeology will permit it to be dated. The earliest stage of construction, Stonehenge I, took place about 2800 B.C., nearly 5,000 years ago. It was built by Neolithic, or New Stone Age, people and, surprisingly, a sophisticated astronomical knowledge is embodied in the earliest design. The linteled circle of so-called sarsen sandstone uprights and central trilithons, or archways, of the same sandstone represent the last stage, or Stonehenge III, and were erected about 2075 B.C. By this time the use of bronze had been introduced into Britain.

(Sarsen sandstone may derive its name from "Saracen" and suggests a foreign origin. The stone is local and was probably hauled overland to Stonehenge.)

Whether we consider the first or last stages of Stonehenge, we are dealing with people who had no known written language. The absence of a historical record has permitted the growth of fanciful chronicles of the stones and at the same time has heightened their mystery.

FIRST STEPS TOWARD THE GIANTS' DANCE

For centuries Geoffrey of Monmouth was regarded as the chief authority on the origin of Stonehenge. From internal evidence in his *History of the Kings of Britain*, it has been concluded that Geoffrey of Monmouth was probably a Welshman and a monk. It also appears that, at least for a time, Geoffrey resided in Oxford. It is possible that there, in

this scholarly environment, during the twelfth century, Geoffrey wrote
his history of Britain. It was finished in A.D. 1136 and is really a pseudo
history. Much of this twelfth-century narrative is concerned with such
legendary figures as Arthur, King Lear, and Brutus, and Stonehenge is
also discussed at length under the name "the Giants' Dance."

According to Geoffrey, Merlin, the magician of Arthurian fame, was
responsible for Stonehenge. In response to a commission from Aurelius
Ambrosius, the British king, to establish a monument in memory of the
460 noble Britons who had been treacherously slain in a sixth-century
battle with the Saxons, Merlin suggests the theft of a ring of immense
stones, called the Giants' Dance, from the Irish. Fifteen thousand men
sail to Ireland and successfully battle the Irish for the ring. The Britons
find themselves, despite their military success, unable to budge the
stones even a Megalithic inch. Merlin intervenes, not with sorcery but
by placing

> in position all of the gear which he considered
> necessary and dismantled the stones more easily
> than you could ever believe.

The stones were put on ships and carried back to England, where
Merlin re-erected them as they had been in Ireland. Geoffrey also tells
that, later, Aurelius, Uther Pendragon (the father of King Arthur), and
Constantine were all eventually buried within Stonehenge, as the monument was called in English in the time of the Saxons.

Henry of Huntingdon, writing perhaps ten years before Geoffrey of
Monmouth, made the earliest known reference to the name Stonehenge and explained that the name means "hanging stones." The name
aptly describes the unique feature of the linteled archways which make
it seem as if the stones are hanging in air. Although approximately nine
hundred stone rings remain in Britain, Stonehenge is the only one to
have lintels.

There may be still earlier references to Stonehenge. Diodorus Siculus,
a Greek geographer who lived during the first century A.D., wrote of a
spherical temple of the sun god Apollo in Hyperborea, which, in all
likelihood, is Britain. The author of the official Stonehenge guidebook,
R. S. Newall, following W. J. Harrison's reference, called attention to
this possible description of Stonehenge by a contemporary of Julius
Caesar and Augustus. The suggestion here of an astronomical connection with the sun may be particularly significant in light of recent astronomical interpretations of Stonehenge.

Although documented speculations of an astronomical use of Stonehenge date from William Stukeley, the English antiquary, in 1740, the

evidence on which such an idea must be based, namely, an accurate survey of the site and its alignments, was not available at all until Sir Norman Lockyer checked some alignments in 1901. Until the most recent
surveys of Alexander Thom, the only other surveys suitable for testing
astronomical alignment were those of C. A. Newham, an amateur
archaeologist, who undertook a study of Stonehenge to satisfy his own
curiosity and interest.

Newham and his wife lived in Yorkshire and visited Stonehenge in
1957. At that time, through a discussion with one of the wardens,
Newham became interested in Lockyer's earlier studies of the monument's axis and alignments. Newham was advised to read R. J. C.
Atkinson's book *Stonehenge*, for Atkinson was the foremost archaeological authority on Stonehenge, and his studies led him to doubt the
validity of Lockyer's work. By the following summer Newham's wife
had died, and he himself was due to retire. Depressed, he decided therefore to return to Stonehenge to survey the site and to investigate the
problem further. Ultimately Newham, who was an amateur astronomer, found a well-defined set of alignments on the sun and the moon
between features of Stonehenge I, the earliest stage of the monument.

Newham's results were first reported in a newspaper in the spring of
1963, half a year before a similar set of alignments were published by
the astronomer Gerald Hawkins. Newham's astronomical interpretation
was ignored, but Hawkins managed to generate great public interest in
his paper in the October 26, 1963, issue of *Nature*. Hawkins received a
great deal of publicity from the news media, and through subsequent
publication of his book *Stonehenge Decoded*, his astronomical interpretation of Stonehenge was widely popularized and accepted by the
public, if not by the archaeologists.

Hawkins used an electronic computer to demonstrate that his astronomical alignments could not be due to chance. He also proposed a second set of alignments, which involved components of the last stage of
the structure, Stonehenge III. Like Newham's alignments, Hawkins'
alignments indicated significant sunrises, sunsets, moonrises, and moonsets in the cycles of the sun and moon. Hawkins also devised an eclipse
predictor from the Aubrey Circle, a ring of fifty-six holes that were dug
and refilled in the time of Stonehenge I. Although Hawkins' com-
puter-supported theory of Stonehenge was novel and appealing, originally it was not based upon an accurate on-site survey. This in part explains the resistance Hawkins met from the archaeologists.

Before we examine the astronomical potential of Stonehenge, we
should review what we know of the prehistoric people who built it. For
this we must rely upon the meticulous work of the archaeologists, in

particular, Stuart Piggott and R. J. C. Atkinson, who carried out careful excavations in the 1950s. Many of their results were published in Atkinson's book *Stonehenge*.

THE PREHISTORY OF STONEHENGE

Prior to 4500 B.C. Southern England was occupied by scattered bands of roving hunters. These people had no agriculture and no domesticated animals, but they worked flint and deer antler with skill and fabricated and used a wide variety of tools. Around 4300 B.C. Neolithic farmers moved into Britain from the European continent and brought agriculture, cattle, primitive wheat, flint and bone tools, and pottery with them. These immigrants are known either as Early Neolithic People or the Windmill Hill People. Windmill Hill, near Avebury, is typical of the "causewayed camps" these people constructed, and many of their known artifacts come from there. They also built long earthen barrows for burial of their dead and cursus monuments, long earth-banked avenues, like that near Stonehenge.

Shortly after the entry into Britain of the Windmill Hill People, other small groups followed them from France. These various Neolithic groups merged with each other and with the original aboriginal inhabitants to produce a so-called Late Neolithic culture. The later-arriving Neolithic people introduced the construction of stone chambers for the group burials under the long earth barrows; West Kennet Long Barrow is the largest, and perhaps earliest, example. These are the first people to work with megaliths, or great stones.

The Late Neolithic culture was still at least a partly nomadic society and was responsible for the construction of Woodhenge, Silbury Hill, the first stage of a site known as the Sanctuary, and Stonehenge I. This comes as a surprise when we remind ourselves of the sophisticated engineering and expert geometry evidenced at Woodhenge and Silbury Hill. It is even more amazing, as we shall discover, because most of the astronomical alignments of Stonehenge are contained in the elements of its earliest stages of construction.

After several centuries of Late Neolithic activity, in roughly 2500 B.C., the Beaker People, from Holland and the Rhineland moved into Britain and introduced the use of worked metal to the islands. The Beaker People were not particularly numerous, but they were successful economically and dominated the culture of Southern England. The name by which they called themselves is as unknown to us as the names of the Neolithic peoples who preceded them, and we have named them for the elaborate clay beakers they so often buried with their dead. It is speculated by some that their economic success was as much due to their introduction of beer into the British Isles as it was to their trade in metals, hence the ubiquitous beakers.

The Beaker People and their Late Neolithic coinhabitants may have jointly built Avebury, the second stage of the Sanctuary, and Stonehenge II. Since we now know that the bluestones which were used in Stonehenge II came from Wales, we might conclude that at this early date these people already had considerable expertise in navigation to travel the ocean waters around Britain. The most sensible transport route from Wales to the site of Stonehenge is by water. It is possible, however, that the bluestones were deposited by natural glacial action. In any case the migrations and trade that were so characteristic of this era may well have stimulated navigation.

As more Beaker People came into Britain, the various cultures blended to form a new one, the Wessex culture. By 1800 B.C. Britain had a Bronze Age culture which was dominated by a wealthy, powerful aristocracy whose trade extended to Central Europe, Ireland, Crete, and Greece. This commercial aristocracy specialized in single burials marked by the prominent barrows that literally cover the landscape of Southern England.

The Wessex culture was probably responsible for the building of the last stages of Stonehenge: IIIa, IIIb, and IIIc. These involved the rearrangement of the bluestones and the transport and erection of the immense sarsen stones. Whatever the methods and motivations of these people may have been, they were building on a Megalithic tradition more than a thousand years old. Yet at Stonehenge we see the introduction of the "hanging" lintels as an innovation peculiar to the Wessex people.

A REVERSAL IN TIME

Before 1901 it was impossible to know how really old Stonehenge was. In that year, however, Professor W. Gowland made several excavations and found, below archaeologically datable layers, about eighty "flint axes," as he called them. Today these are known to be cores from which flakes were struck and as such are not datable, but Gowland asserted that these showed that the monument was definitely a product of the Stone Age. A trace of copper at the base of a sarsen stone, probably from a metal tool, showed that copper was coming into use.

Archaeological investigations had been carried out as early as a century before Gowland's by William Cunnington, but without result. He failed to find anything he could use to date the monument, but he very considerately buried a bottle of port wine underneath the so-called

Slaughter Stone for the benefit of future investigators. We know Cunnington buried the bottle because it was found in 1920. Sadly, the seal had broken, and the cork had disintegrated.

A radiocarbon date for Stonehenge became possible in the 1950s when Atkinson, Piggott, and J. F. S. Stone found charcoal and other artifacts in some of the holes which comprise the large circle outside the stones of the monument. This ring and the holes are named after the British antiquary John Aubrey, who discovered them in the seventeenth century. The radiocarbon dating method indicated a preliminary date of about 1850 B.C. This date had much to do with the subsequent interpretation of the place of Megalithic monuments in prehistory.

Stone monuments are found in many parts of Western Europe. The earliest known examples of stone architecture are found on Malta, in the Mediterranean. When a radiocarbon date for Stonehenge became available, it was also thought that the design of Megalithic tombs derived from Mycenaean *tholos* tombs in Greece. Similarities between the two types of structures were cited. The cruder construction of the megaliths was interpreted as the product of a less-advanced, Western European culture which had adopted the style of its more sophisticated Mediterranean neighbors. A pattern of cultural diffusion from the Mediterranean, and ultimately from the Near East, was thought to have guided the rest of Europe out of its primitive state and on to civilization.

Since the 1950s, more archaeological evidence was compiled, and it seemed to suggest that something was wrong with the picture of cultural diffusion. Discrepancies between radiocarbon dates and historical dates for sites in Egypt were surfacing. Eventually the radioactive carbon dating scheme was recalibrated against tree ring counts of the bristlecone pine. These pines are the oldest known living things and are found in the White Mountains of California, overlooking Owens Valley.

A radiocarbon date is obtained by measuring the amount of radioactivity left in a sample of organic material. All living things contain carbon, and a very small fraction of these carbon atoms are radioactive. They decay, and the extremely constant rate of decay permits us to estimate how much time has elapsed since the once-living matter died. A tree, an animal, or any other living thing no longer incorporates carbon into its system after it dies. The radioactive carbon continues to decay, and we can measure today, by the amount of radioactivity still present in the sample, how long it has been since the living thing died. The accuracy depends upon the sample and its age, but in general radiocarbon dates are estimated as good to ±100 years. This simple system therefore permits a direct access to the dates of prehistoric and ancient material.

Radiocarbon dating is dependent upon a number of assumptions, however, and one of these was that the amount of radioactive carbon-14 in the atmosphere has been the same for the last several thousand years. We now know that this is not the case. Samples of bristlecone pine have been independently dated by counts of their rings. These dates are systematically older than the radiocarbon dates of the same samples of wood. They show that the amount of radioactive carbon has varied significantly. This variation is linked in turn to the number of primary cosmic ray particles that fall on the earth's upper atmosphere, for these particles are responsible for the production of the carbon-14.

Instead of a steady rain of cosmic ray particles, we now imagine periods of light drizzle and heavy showers. Part of this variation is caused by cyclical changes in the strength of the earth's magnetic field, which controls the activity of the charged particles. Another variation is caused by the sources of the particles. Careful analysis has resolved individual periods of cosmic ray shower intensity, which can be related to specific supernovae, or exploding stars, not too distant from the sun.

In general, radiocarbon dates for European prehistory are systematically too recent, and in fact, the older the radiocarbon date, the more in error it is. When corrections were applied to the radiocarbon dates of Megalithic monuments, the tombs were found to be older, and not younger, than the Mycenaean *tholos* tombs. The pattern of diffusion from the Mediterranean to Northern and Western Europe had to be abandoned. The Megalith Builders are now recognized as the independent inventors of an impressive pre-Mediterranean culture. The earliest stage of Stonehenge is as old or older than the pyramids of Egypt. The British Museum Radiocarbon Laboratory recently obtained a new date for the Avenue and Stonehenge II. Calibrated by bristlecone pine, it is 2130 B.C. Stonehenge III was constructed near 2075 B.C. These dates should end once and for all the popular speculation that Stonehenge was conceived, designed, and executed by a traveling merchant-architect from Greece. No longer can we suggest that this Mediterranean genius lies buried under Silbury Hill. Indeed, no one has been found buried within Silbury Hill. A link to the Mediterranean was drawn by the discovery of the carving of the Mycenaean dagger on one of the sarsen stones at Stonehenge. With bristlecone hindsight, we now must conclude that either the dagger is not Mycenaean after all or that it was inscribed into the monument long after it was built, much as Vikings carved runes into the walls of Maes Howe, the huge Neolithic burial mound in the Orkneys.

The carvings of daggers on one of the sarsen stones suggested a Mediterranean influence on the design and construction of Stonehenge until its date was recalibrated through studies of the bristlecone pine. (E. C. Krupp)

A "SPHERICAL" TEMPLE

We have already seen that the Greek Diodorus Siculus, who wrote circa 8 B.C. of the northern island inhabited by Hyperboreans, may have been referring to Stonehenge when he referred to a spherical temple of Apollo that was found there. Newall, who exhumed this quotation and called attention to its possible significance, also reminds us that classical writers often used the term "spherical" to mean "astronomical," and in that case we interpret Diodorus' words as a description of the use, and not the shape, of the temple "beyond the north wind." We may therefore have a record, admittedly vague, of considerable antiquity of an astronomical structure in Britain. The first century B.C. is, of course, too recent to allow certain knowledge of the Stonehenge of the third millenium B.C., but Diodorus relied upon Hecataeus of Abdera, who flourished about 300 B.C., and it may be that intervening peoples continued to appropriate and use the "temple" long after its completion.

The first concrete references we find for an astronomical use of Stonehenge are from the eighteenth century A.D. William Stukeley, the author of *Stonehenge, A Temple Restored to the British Druids* (1740), mentioned that the monument's axis is aligned with the summer solstice sunrise. This is not the same line that connects the center of the ring to the Heel Stone, however. The Heel Stone is a sarsen boulder located about 256 feet northeast of the center. It towers 16 feet above the ground, and its weight is estimated at 35 tons. Many explanations of its name have been offered, but it seems most likely to be a product of confusion and misidentification. In Newall's *Stonehenge Wiltshire* (the official British Government guidebook), Dr. John Smith is credited as the first, in 1771, to propose that the Heel Stone was the solstitial sunrise indicator. Dr. Smith may also have been the first to mention the possibility of a distant foresight, for he also describes the sunrise in relation to a hill on the northeast horizon.

Other solstitial risings and settings were suggested in connection with several points around the monument, in particular, the Station Stones and Station Mounds. These nearly form a rectangle, the corners of which fall upon the Aubrey Circle. The short sides of the figure are aligned northeast–southwest and could, as the Reverend Edward Duke said in 1846, indicate summer solstice sunrise and winter solstice sunset. No solid data were available for any of these conclusions until 1901, however, when Sir Norman Lockyer actually surveyed the monument.

Lockyer surveyed other sites in Britain besides Stonehenge, from Cornwall to Orkney, and found astronomical indicators in many Megalithic constructions. He concluded that the axis, or Avenue, of Stonehenge was aligned on the summer solstice sunrise and tried to derive a date for construction based upon the alignment. Lockyer's date was calculated to be somewhere between 1880 and 1480 B.C. Assuming that the prehistoric astronomers of Stonehenge observed sunrise alignments at the first gleam and sunset alignments at the last gleam, Lockyer also noted a sunset alignment for the May Day–Lammas declination. The reverse direction, he claimed, marked sunrise for Martinmas–Candlemas. These calendrical alignments and Lockyer's dating have been challenged since their enunciation and are not regarded seriously today.

Lockyer also reaffirmed the opinion of Sir Flinders Petrie, the renowned archaeologist, that the Heel Stone, as viewed from the monument's center, marked the sunrise position at the summer solstice, and this astronomical association is the one most familiar to the average visitor to Stonehenge. There was little other popular or professional interest in the astronomical alignments of Stonehenge until 1963. Modern Druids continued to show up in the chilly predawn of the summer solstice, but they were accompanied more often than not by crowds more interested in ale and mischief than alignments and mysteries.

THE "DECODING" OF STONEHENGE

In 1963 Gerald Hawkins, a British-born astronomer and at that time at the Smithsonian Astrophysical Observatory in Massachusetts, published a letter in the British journal *Nature* asserting twenty-four alignments and indicated directions of astronomical importance at Stonehenge. A second letter from him quickly followed, and it interpreted some of Stonehenge's features as a computer of eclipses. Finally, six months later, a third discussion appeared in the American journal *Science*. In it Hawkins discussed astronomical alignments at Callanish in Scotland's Outer Hebrides. This effort was primarily designed to show that Stonehenge was not unique.

When Hawkins' book *Stonehenge Decoded* was published in 1965, its rather presumptuous title and Hawkins' use of a digital computer captured the public imagination. Mentions of Stonehenge found their way into the introductory sections of astronomy textbooks, and Stonehenge shows became stock-in-trade at planetaria.

Hawkins divided his alignments into two groups. The first of these included alignments between features of the earliest stage of construction, Stonehenge I. The first elements of Stonehenge were the circular bank that surrounds the entire monument, the Aubrey Circle inside the bank, the four stations, the Heel Stone, and miscellaneous post and stone holes found in various places near the monument. The stations are four points on the Aubrey Circle. Two are marked by mounds and two are marked by stones. Together they comprise the four corners of a nearly perfect rectangle. The stations are identified on the official plan as points 91, 92, 93, and 94. The bank was constructed from chalk rubble dug from the ditch just outside of it. It is roughly 320 feet in diameter and originally stood 6 feet high. An entrance gap, at least 40 feet in width, was left on the northeast side when the bank was constructed.

The Aubrey Circle is a ring of fifty-six holes which were refilled after being dug. The holes never contained wooden posts nor stones but were filled with chalk, cremation remains, and a few bones. The circle has a diameter of 284½ feet, and it was named after John Aubrey by

STONEHENGE

Professor Alexander Thom and his associates have completed this new plan of most of the features of Stonehenge. This plan is accurate enough to permit careful examinations of possible Stonehenge astronomy and geometry, and many of the components of Stonehenge described in the text can be located on this plan in order to understand Stonehenge's construction and design. (Griffith Observatory, after Alexander Thom)

Stonehenge I Alignments (Hawkins)

Gerald Hawkins identified numerous astronomical alignments between the features of the earliest phase, Stonehenge I. Some solstitial alignments (+24 and −24) and lunar standstill alignments (+19, −19, +29, and −29) appeared to fit the rectangular geometry of the stations, while others seemed to avoid any obvious geometric pattern. (Griffith Observatory, after Gerald S. Hawkins)

Newall. Aubrey, who discovered the holes, was also the first to claim that Stonehenge was built and used by Druids.

A series of post holes is located in the Causeway, just outside the entrance gap. Another group of Stonehenge I post holes is found to the southeast, between the large sarsen stone circle and the Aubrey Circle. Stone holes B and C were found in the Avenue, and stone hole E is situated in the entrance gap. There are also the four post holes, known as the A post holes, in the Avenue, slightly in front of and to the northwest of the Heel Stone. In 1966 three tree-trunk-sized post holes were discovered about 830 feet northwest of the center of Stonehenge in the course of excavation of an extension to the car park. Most of these holes are unmarked and unnoticed.

Finally, small circular ditches marked Station 92 and Station 94. Sometime later, stones were probably placed at these positions and at Stations 91 and 93, where stones remain today. The stations nearly mark the corners of a rectangle and with the Heel Stone embrace most of Hawkins' Stonehenge I alignments.

Hawkins reiterated the now-familiar theme of solstitial alignment of the Heel Stone. Several other solar alignments were included in the list in *Stonehenge Decoded* also, including the following:

ALIGNMENT	ASTRONOMICAL PHENOMENON
Center to Heel Stone	summer solstice sunrise
93 to 94	summer solstice sunrise
92 to 91	summer solstice sunrise
G to 94	summer solstice sunset
94 to G	winter solstice sunrise
93 to H	winter solstice sunrise
91 to 92	winter solstice sunset
94 to 93	winter solstice sunset

Just as the sun would reach its highest altitudes in summer and dominate the daytime sky, the winter full moon would light the long, hyperborean nights and trace its highest arc through the sky. In addition to the solar alignments, Hawkins listed a series of lunar alignments, at major and minor standstills and primarily for the northern moon (a few southern standstill alignments were also included):

ALIGNMENT	ASTRONOMICAL PHENOMENON
92 to G	northern major standstill moonrise
Center to A	northern major standstill moonrise
Center to D	northern major standstill moonrise
91 to 94	northern major standstill moonset
Center to F	northern minor standstill moonrise
Center to 93 (or 91 to 93)	northern minor standstill moonset
93 to 92	southern major standstill moonrise
Center to 91 (or 93 to 91)	southern minor standstill moonrise

Six out of eight possible standstill alignments are present. No southern standstill moonsets are indicated. Many of the alignments are reversals of a line with another astronomical significance. The diagonal between 93 and 91 may have been originally in use, but the line of sight is now blocked by the massive stones of Stonehenge III.

It is the components of Stonehenge III, the Sarsen Circle of evenly spaced uprights laid out on a ring 100 feet in diameter and the Trilithon Horseshoe of five towering stone gateways, that first come to mind when we think of Stonehenge. Stonehenge III was the last ver-

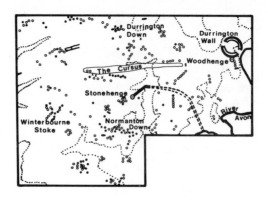

The Stonehenge
Region

In the region of Stonehenge the path of the Avenue can be traced from the nearest bend of the River Avon and over the easiest grade to the position of Stonehenge. (Griffith Observatory, after J. F. S. Stone)

sion of the structure, however, and it also involved some modification of the previous stage, Stonehenge II.

Stonehenge II incorporated the Avenue, which consists of two parallel chalk banks 47 feet apart. These run northeast from the Causeway about one third of a mile. The Avenue can be traced down into the valley now occupied by farmland. There it turns nearly due east to the crest of a hill, where it turns again, to the southeast, and runs another half mile to the banks of the Wiltshire Avon. This peculiar route was not discerned in its entirety until aerial photography revealed it in 1923. The motivation for its strangely directed course became clear when it was realized that the Avenue traces the path of easiest grade from the river to the site of Stonehenge. It very likely marks the path along which the bluestones were dragged in order set up the double circle of Stonehenge II. The Avenue joins the Wiltshire Avon at the point where the river most closely approaches Stonehenge.

The large sarsen stones of Stonehenge III originate from the Marlborough Downs, about twenty miles north of Stonehenge, where similar sandstone boulders can still be seen. The bluestones, by contrast, are spotted dolerite, an igneous rock, and have been proved to derive from the Preseli Mountains in Pembrokeshire, Wales. In fact, bluestone boulders of every size and shape may be seen there today. It has been suggested that the bluestones were transported overland from Preseli to Milford Haven on the coast and then moved by sea up the Bristol Avon. A series of river portages may then have completed the 240 mile journey to the Salisbury Plain.

Wales is a long way to the west to go for stone that might just as easily be found nearer to home. There was apparently something special about the bluestones that motivated their use. Although Wales is not as far west as Ireland, Geoffrey's chronicle may be echoing an ancient and garbled tradition. Merlin emphasized that the stones had special properties. The name "bluestone" is given because the stones are blue-green in color. The color is most obvious when the stones are wet, and the white spots on the spotted dolerite also stand out best against the dark background of blue then. Whatever the meaning of the bluestones may be, they were erected in a double, incomplete ring. The concentric circles were revealed only by the Q and R holes, which were found during excavation.

It is unclear what the real intention of Stonehenge II was, for the two bluestone circles were left with a wide gap in the west. Although some have interpreted these two bluestone crescents as an abortive design, the plan might just as well have been intentional. An entrance-way of bluestones was constructed on the northeast side, aligned with a large pit located opposite the entrance, on the southwest side of the double ring.

The third stage of Stonehenge began with the dismantling of original bluestone rings. A new ring of larger, sarsen stones, capped with lintels, was constructed. This Sarsen Circle originally was composed of thirty uprights which were sized and dressed. The thirty lintels were cut to curve with the circle and were held in place with mortise-and-tenon and tongue-and-groove construction. As an additional precaution against slippage, the uprights and the lintels were dished and chamfered.

Remaining uprights in the Sarsen Circle weigh about 25 tons each and stand 14 feet above the ground. Between 3 and 7 feet of each stone are buried in ground. The lintels weigh 7 tons each. Five taller trilithons were erected to form a horseshoe inside the Sarsen Circle. The trilithon uprights are the largest in the monument, and those that formed the Southwest Trilithon, at the bow of the horseshoe, are the largest of all. This trilithon stood 24 feet above ground level, and its western upright weighs 50 tons. The axis of the horseshoe is aligned with the northeast Sarsen Circle archway and with the Avenue.

At the same time the major features of Stonehenge III were put into place, two large stones were set upright as a gateway on the axis, about halfway between the Sarsen Circle and the Heel Stone. One of the stones has disappeared, but the second is the Slaughter Stone, under which Cunnington so thoughtfully left the bottle of wine. The name of the Slaughter Stone is the product of the overworked imagination of Druidophiles, and there is no evidence to justify its bloody reputation.

The builders of Stonehenge III also intended to reuse the bluestones and set some of them into place in an oval pattern inside the Trilithon Horseshoe. They also dug the two rings of Y and Z holes outside the Sarsen Circle, presumably to hold the remaining leftover bluestones. For some reason the builders abandoned their plans and removed the bluestones they had placed in the center. A horseshoe of bluestones was then set up inside the Trilithon Horseshoe, and a circle of bluestones was erected between the Sarsen Circle and the Trilithon Horseshoe.

Hawkins' second group of astronomical alignments is really a set of indicated directions which appear to have been built into the structures of Stonehenge III. In almost every case the view is framed by the uprights of a trilithon and the uprights of a Sarsen Circle archway.

SARSEN ARCHWAY	
INDICATED DIRECTIONS	ASTRONOMICAL PHENOMENA
30-1 to Heel Stone	summer solstice sunrise
59-60 to 23-24	summer solstice sunset
51-52 to 6-7	winter solstice sunrise
55-56 to 15-16	winter solstice sunset

Just as Hawkins had noted solar and lunar alignments for Stonehenge I, he listed directions for the sun and the moon for Stonehenge III. The solar phenomena above are a complete set, but only half of the possible lunar standstills appear to be present.

SARSEN ARCHWAY	
INDICATED DIRECTIONS	ASTRONOMICAL PHENOMENA
57-58 to 21-22	northern major standstill moonset
57-58 to 20-21	northern minor standstill moonset
53-54 to 9-10	southern major standstill moonrise
53-54 to 8-9	southern minor standstill moonrise

A glance at a plan of Stonehenge will show that three of these indicated directions are based on incomplete pairs.

Later, in *Stonehenge Decoded* Gerald Hawkins added eight more Stonehenge I alignments to his original sixteen. All of these involved the sun or the moon at the equinoxes, and Hawkins said that C. A. Newham had given him the clue to them. Newham had been working independently on Stonehenge astronomical alignments for many years.

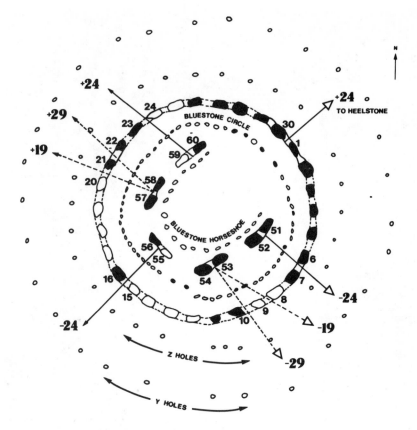

Stonehenge III Alignments (Hawkins)

Hawkins also discovered several astronomically indicated directions through gaps in the uprights of the trilithons and of the Sarsen Circle. Again, solstitial (±24) and lunar standstill (±19 and ±29) phenomena appeared to have been intended. These alignments involve components of Stonehenge III. (Griffith Observatory, after Gerald S. Hawkins)

Some of the Stonehenge III alignments are really rather wide. The view through stones 57–58 and 21–22 includes the major standstill northern moonset. (E. C. Krupp)

His smaller, more reliable, and more plausible set of alignments had been reported in the *Yorkshire Post* seven months before Hawkins' letter appeared in *Nature*. Newham had clearly seen the equinoctial alignment that Hawkins had missed on his first time around.

IN THE DARK OF THE MOON

Hawkins did not restrict his exploration of Stonehenge astronomy to lunisolar alignments. He also immersed himself in the perilous waters of eclipse prediction. The tantalizing commentary of Diodorus Siculus might well tempt even a conservative interpreter, however.

Diodorus reported that the moon, as viewed from Hyperborea, "is seen only a little above the earth and has certain definite prominences on its surface, just like the earth." In the next breath Diodorus mentions that the god, presumably the moon, "returns to the island every nineteen years, the period when the stars complete their cycle." Diodorus referred to this nineteen-year period as the "year of Meton" after the fifth-century B.C. Greek astronomer credited with the cycle's discovery, and it would be reasonable to conclude his reference is a garbled description of the Metonic cycle. This cycle, nineteen years in duration, is the sequence of lunar phases relative to the solar calendar. If there were a whole number of full moons in each 365.25 days, the same phases of the moon would occur on the same solar calendar dates each year. Instead, every nineteen years the same lunar phases occur on the same calendar dates.

THE METONIC CYCLE

235 synodic months $\times$ 29.5306 days per month $=$ 19 tropical years $\times$ 365.2425 days per year

6939.69 days $=$ 6939.61 days

The two intervals are nearly equal and permit one to predict the phase of the moon on a particular date.

In seeking a better understanding of the reference to nineteen years, Newall wrote to Hawkins and asked him if the full moon might do "something spectacular" once every nineteen years at Stonehenge. Stimulated by this inquiry, Hawkins responded that a total lunar eclipse would be just the sort of spectacle needed. He carried his hypothesis one precarious step further and judged that an eclipse of the moon visible over the Heel Stone or through the Southwest Trilithon would be the most spectacular of eclipsed moons.

After calculating the particulars for lunar eclipses between 200 and 1000 B.C., Hawkins concluded that an eclipse of the moon or sun always occurred when the winter solstice full moon rose over the Heel Stone. Half of the lunar eclipses would have been visible from Stonehenge.

If the winter full moon rose over D or F, as viewed from the center, it would be the season of major or minor standstill, respectively. In the fall, six months later, the full moon would be at one of the nodes, and an eclipse might be visible. By contrast, the winter solstice full moon rises over the Heel Stone, midway between D and F, midway through the period from major standstill to minor standstill. In any case, a complete cycle, which would bring a midstandstill lunar eclipse back to the same calendar date, would occur every 18.61 years. This is not an easy number to use in a counting cycle, as Hawkins pointed out, but he countered that a better approximation and counting scheme could be obtained by working with three cycles, for

$$3 \text{ cycles} \times 18.61 \text{ years per cycle} = 19 \text{ years} + 19 \text{ years} + 18 \text{ years}$$
$$55.83 \text{ years} = 56 \text{ years}.$$

Therefore, the number 56 could refer to such a counting sequence. The appearance of an astronomical association with the number 56 led Hawkins back to the mystery of the Aubrey Holes. They could be used to count years and indicate eclipses over three complete regressions of the moon's orbit.

Hawkins began his description of his Aubrey Hole eclipse predictor at a winter solstice lunar eclipse. This could occur only when the full moon is on a node and therefore when it has risen above the Heel Stone. In that year a white stone marker is located on Aubrey Hole 56, which is roughly on the axis of Stonehenge. A total of six stone markers are employed, three black and three white. The six counters alternate in color around the Aubrey Circle and are spaced in a sequence of nine, nine, ten, nine, nine, ten Aubrey Holes. In the following year each stone is moved one hole, counterclockwise. The white stone of interest is now on Aubrey Hole 55, a so-called "safe" hole. No winter solstice eclipse occurs. An eclipse may occur, but it will not be accompanied by the full moon rising over the Heel Stone. Hawkins did not mention this possibility and concluded that "nothing spectacular happens."

All stones are moved one hole each year. Five years after the initial eclipse the same white stone under consideration occupies Aubrey Hole 51. The full moons at the vernal and autumnal equinoxes are now at the nodes. An eclipse is imminent.

In another four years Aubrey Hole 56 is occupied by a black marker. An eclipse is again possible. The system keeps running with markers at 51 and 56 predicting lunar eclipses in the years of major standstill (equinox eclipses) and in the years of midstandstill (solstice eclipses),

Hawkins Eclipse Predictor

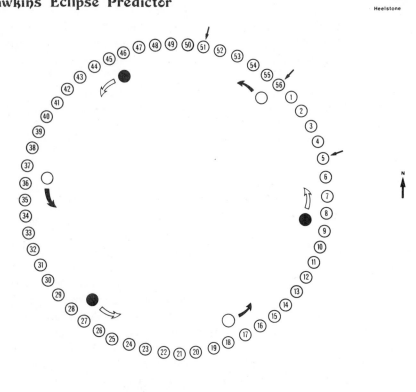

One variation of the Hawkins eclipse predictor involved six alternately white and black stones spaced at intervals of nine, nine, ten, nine, nine, ten holes around Aubrey Circle, shown in the diagram. Each stone is moved one hole each year. If a stone should land on a "special" hole, a certain kind of eclipse may occur. (Griffith Observatory, after Gerald S. Hawkins)

respectively. Similarly, Aubrey Hole 5 predicts eclipses in the years of minor standstill (equinox eclipses).

What seems odd about the Aubrey Hole eclipse predictor is that the alignments are already there to tell the Stonehenge astronomers when the standstills take place. That in itself is sufficient information to tell them what kind of an eclipse can occur that year. The Aubrey Circle in this guise is redundant. Even more unsatisfying is the fact that only a small number of all potentially visible lunar eclipses are predicted by this "computer."

Hawkins later streamlined his eclipse predictor down to three stones and used Aubrey Holes 28 and 56 to predict solstice eclipses. Errors in the averaging of cycles will eventually add up to a noticeable discrepancy. When this occurs, the markers can be reset to recalibrate the prediction with the proper year. All these considerations aside, we still find the initial question, "What spectacular thing could the moon do every nineteen years?" unanswered. The full moon will rise over the Heel Stone at the winter solstice every 9.3 years.

THE ARCHAEOLOGISTS STRIKE BACK

The Hawkins interpretation of Stonehenge immediately generated what English archaeologist Colin Renfrew has called "one of those agreeably fiery little controversies to which archaeology seems particularly prone." In a highly critical article in *Antiquity*, entitled "Moonshine on Stonehenge," R. J. C. Atkinson, the foremost Stonehenge archaeologist and author of the definitive book on the subject, skeptically reviewed *Stonehenge Decoded*. In a shorter review in *Nature* Atkinson charged that Hawkins' book was "tendentious, arrogant, slipshod, and unconvincing." Atkinson's criticism was directed against the archaeological content of Hawkins' book. With the archaeology in question, a number of respectable archaeologists naturally were dubious about the book's astronomical content. A series of complaints were aired.

It was said that Hawkins disregarded changes in the ground level due to erosion since 2000 B.C. About 1½ feet of earth are estimated to have worn away. Hawkins later claimed that tool marks on the stones indicate the original ground level and that recent deposition of gravel brought the ground back to its original level. He participated in a photogrammetric survey for good measure, however, included the effect of the sloping terrain, and reduced his proposed astronomical alignments by a few.

Objections were raised against the trilithon-sarsen archway alignments. These were alleged not to be geometrically exact, and the great width of the field of view permits considerable leeway in declination. Also, Hawkins has never been particularly concerned about interpretation of the uncertainties of the alignments. His original errors in alignment were typically 1 to 1.5 degrees from the expected positions of the sun and the moon. These errors are considerably larger than those determined by Alexander Thom for other Megalithic sites and highlight the difference in approach of the two methods. An error of 1.5 degrees is roughly equivalent to missing the mark by three lunar or solar diameters. Hawkins is satisfied with the trilithon-sarsen archway combinations because his interpretation of Stonehenge does not require the monument to be a genuine precision astronomical observatory. What Hawkins is saying is that the features of Stonehenge III could have been used in the fashion he describes. Despite his colorful descriptions, the hypothesis is not the same as proof, and the door remains open for alternative explanations.

The original data used by Hawkins were criticized as inadequate, for they had been extracted from a plan of Stonehenge that was not free of error. Hawkins had not used reliable on-site survey data of his own but had appropriated inaccurate data from an unsatisfactory source. As mentioned already, Hawkins later carried out a photogrammetric survey and confirmed his earlier results, within the restrictions, however, of his approach.

ACCORDING TO HOYLE

Glyn Daniel, the editor of *Antiquity*, the British archaeological journal, invited the well-known and highly respected British astronomer and cosmologist Fred Hoyle to examine critically the Hawkins theory from the astronomical point of view. Hoyle did that and went the archaeologists one better by devising his own astronomical scheme for Stonehenge.

From the start, Hoyle realized a computer was not needed to carry out Hawkins' analysis, but Hoyle felt that the image of Stonehenge as an astronomical device was right. Basically, Hoyle confirmed Hawkins' results. The statistical validity of the Stonehenge I alignments seemed real. Hoyle estimated that the sizes and distances between stones permitted angular measure precise to nearly $\pm\frac{1}{4}$ degree. Because the practical needs of a pastoral community do not require this precision, Hoyle discarded a calendrical interpretation of Stonehenge, just as Hawkins had.

The lunar indications suggested a prehistoric knowledge of the 18.61 year cycle of regression of the line of nodes. To Hoyle this meant that the Stonehengers knew what they were doing when they laid out and constructed the circle of Aubrey Holes.

With justification, Atkinson had pointed out that a circle of fifty-six evenly spaced holes and with a diameter of 284 feet 6 inches was hardly necessary to count the years in three lunar cycles. Hawkins gamely countered that his was the first and *only* explanation of the Aubrey Circle and implied that if the archaeologists were so unhappy with his theory, they could go out and explain the number 56 themselves. This time Hoyle sided with the archaeologists. He also developed his own remarkable interpretation of the Aubrey Holes.

A gnawing doubt lingered. Did the Hawkins eclipse predictor really work? Hoyle instinctively sensed the observational approach of the Stonehengers. Alexander Thom's work at hundreds of other Megalithic sites confirmed the truth of this notion even more strongly than Hoyle could have known from Hawkins' Stonehenge I alignments. The Hawkins eclipse predictor seemed more a product of numerological astronomy, like that of the Babylonians, than something that would emerge from the practical, observational approach of the Megalith Builders. Worse still, the Stonehengers would need a table of eclipses to develop the method, just as Hawkins needed one. There is no evidence they made and kept such tables.

Even more strongly Hoyle objected that the Hawkins eclipse predictor could only herald a small fraction of the eclipses, namely, the solstice and equinox eclipses. There might, thought Hoyle, be something to the 56 and 18.61 relationship, and so he tried to incorporate it into his own interpretation.

Hoyle emphasized the importance of the position of the lunar nodes. These, relative to the sun and moon, determine the occurrence of eclipses. The sun and the moon must both occupy a node, or be very close (within 10 degrees) to them, for an eclipse to take place. If the positions of the sun, moon, and both nodes are known, eclipses may be predicted. Knowledge of the calendar date is equivalent to knowledge of the sun's position on the ecliptic. A complete circle could be used to represent the sun's changing position through the year. Each day a sun marker could be moved 360 degrees/365¼ days on the ring. It is possible, of course, to divide the circle into other units and to move the marker some other amount in some other interval of time. In the end, the marker must complete the circle in a year. Hoyle rightly pointed out that 360 is an unlikely number because it represents "bad" astronomy, a poor approximation to the length of the tropical year.

Suppose, on the other hand, the Stonehengers divided the circle into fifty-six parts. A marker which is moved two divisions every thirteen days on a circle divided into fifty-six divisions will complete the year's round in 364 days. An error of 1¼ days, or about 1 degree, results. The error is considerably less than 10 degrees and may be adequate for eclipse prediction. If the sun marker is reset on the solstices every six months, the error can be reduced to ½ degree. This is only possible if the solstices can be observed accurately. Could Stonehenge I do it?

Hoyle estimated the sun could be located no better than 1/250 of its six-month swing across the horizon. Stonehenge I alignments were good to only ±5 days and were, therefore, not so good. Undismayed, Hoyle proposed an alternate method of date determination. The method is basically a simple interpolation. The date of solstice is inferred from observations of the sun on several days before and after the solstice.

An outlying stone that is placed to give a solar alignment 1½ degrees south of the summer solstice sunrise indicates a sunrise that occurs several weeks before and after the solstice. The solstice date would be midway between the two observations of sunrise made with this stone. Counting backward from the time of the second observation one half the number of days in the interval between indicated sunrises could provide the date of solstice. With a similar outlying stone in the southeast, the calendar could be corrected every six months.

Hoyle's method included the definition of sunrise as the first gleam of the sun. This is an easier observation to make than Hawkins' full orb, and it permits greater accuracy. A set of markers for several suitable days might be set up in case bad weather prohibited observation on any one of them. Hoyle interpreted the multiplicity of certain alignments and Hawkins' relatively large errors as evidence of an interpolation procedure. The Stonehengers "errors" were really intentional offsets, for nineteen of Hawkins' alignments were alleged to be off in the correct sense to obtain the true extreme. Hoyle thought that the number of cases was statistically significant, but Atkinson has shown that only fourteen "intentional errors" are present and that the number is too few to be significant. The same theory of intentional "errors" was applied to the four A post holes, which are located just north of the axis of Stonehenge and near the Heel Stone. Hoyle imagined these were used to interpolate the time of northern major standstill of the moon.

Once Hoyle was confident that the dates of astronomical phenomena could be correctly determined, he returned to the problem of eclipse prediction. The second step was inclusion of the moon's motion on the Aubrey Circle. A moon marker can be started by placing it on an Aubrey Hole opposite the sun's marker on the date of full moon. Each day the moon marker must be moved two of the fifty-six divisions, and its position can be corrected by resetting the marker at each new and full moon. Lastly, the node markers must be set. At a major standstill, one node, the ascending node, is 90 degrees clockwise from the summer solstice on the Aubrey Circle. The descending node is opposite the ascending node, and both node markers move three Aubrey Holes each year, but in the opposite direction, counterclockwise, from the motion of the sun and moon. Now with all markers calibrated and moving at their appropriate rates, we neo-Stonehengers need only watch for those

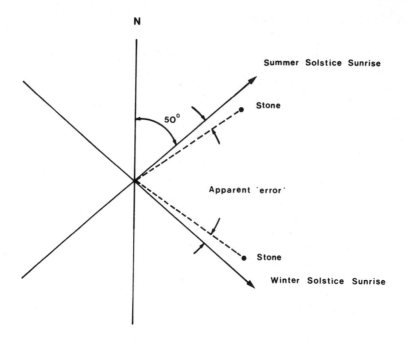

N

Summer Solstice Sunrise

Stone

50°

Apparent 'error'

Stone

Winter Solstice Sunrise

Fred Hoyle suggested that some of the Stonehenge solstice alignments were intentionally offset by the Stonehenge astronomers from the correct directions to permit them accurately to interpolate the exact day of solstice, which would have been otherwise difficult to pinpoint directly. (Griffith Observatory, after Fred Hoyle)

times when the angular separations between sun, moon, and node markers are less than the ecliptic limits for lunar and solar eclipses. Hoyle's device is ingenious, but its design requires a foreknowledge of the number of required markers, of the markers' rates and directions of motion, and of the proper dates for calibration. If the Stonehengers actually did use the Aubrey Circle in this way, they had an ability for abstract scientific thought that exceeds previous assessments.

Hoyle's speculations finally carried him to unprecendented fancies. He assumed that the sun, moon, and lunar node were regarded as gods by the builders of Stonehenge. Presumably they would not distinguish between the ascending and descending nodes. To Hoyle the lunar node would evolve into a powerful, invisible god, a god whose strength permitted the eclipse of the sun and moon. The node is the root, according to Hoyle, of the "invisible and powerful" God of Isaiah. The sun, moon, and node, he says, are the source of the Trinity.

Hoyle Eclipse Predictor

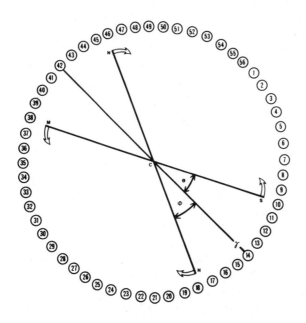

Hoyle also devised an eclipse predictor, but his foretold all potential eclipses. He, too, sent marker stones around the Aubrey Holes, but his markers were not merely counters. They actually represented the sun (S), the moon (M), the two nodes of the moon's orbit (N), and the location of the vernal equinox (γ). Movement around the Aubrey Circle of these markers was equivalent to changes in position in the sky. Each marker was sent at the appropriate rate around the Aubrey Circle to complete the object's cycle in the proper interval of time. The line of nodes, NN, turns clockwise as the node markers are moved. The sun and moon can be started out by setting them opposite each other at the time of full moon. Once this is done, they each proceed around the circle at their own rates. The angles ⊙ and φ indicate the location of the sun and the line of nodes with respect to the vernal equinox. When the sun and the moon land on the same or opposite nodes at the same time, an eclipse may be visible. The Heel Stone and the A post holes are shown, for reference, to the northeast of the Aubrey Circle. (Griffith Observatory, after Fred Hoyle)

AN AMATEUR'S INSIGHT

The series of brief commentaries on Stonehenge astronomy in *Antiquity* unfortunately was summed up by the well-known British archaeologist Jacquetta Hawkes in her contribution, "God in the Machine." Her summary was a fair and informative report, but she concluded that astronomical alignments, even if real, were of small interest. The sensationalism of Hawkins and the fancies of Hoyle may have contributed to archaeologists' reluctance to take Stonehenge astronomy seriously. Had more attention been paid to the relatively unpublicized work of Newham, the outcome might have been different.

After retirement in 1958 from a post with the British gas industry, Newham was looking about for something to do. He was neither a professional astronomer nor archaeologist, but he knew how to survey and he understood practical astronomy. Having read the summer before of Lockyer's unfruitful attempts to interpret Stonehenge astronomically, Newham decided to survey the alignments himself. Early on he found what he thought was a major standstill moonrise line, 92 to G, but he was disappointed and delayed by a mistyped check of the bearing that disagreed with his alignment. The typing error was not discovered until 1962, and on the night Newham found it he also worked out several other astronomical alignments. Newham's findings were based upon on-site data, which he had obtained, and he made his work available to two celebrated scholars of Stonehenge, Newall and Atkinson.

Newall was skeptical of most of Newham's results, but Atkinson was sympathetic and advised publication in *Antiquity*. *Antiquity* declined publication, perhaps because Newham lacked professional credentials. As an alternative, Newham engaged a printer to publish the findings as a booklet in June 1963, but publication was delayed by a completely destructive fire at the printing works. The first galley proofs were not available until late October, three days before Hawkins' Stonehenge letter appeared in *Nature*. A review of Newham's results had been published on March 16, 1963, in the *Yorkshire Post*, but the article went unnoticed. When the Hawkins paper was published, the response of the media was vociferous, and Newham's careful work was temporarily lost in the shuffle.

Newham gradually regained some of the loss of recognition after his pamphlets *The Enigma of Stonehenge* and its *Supplement* were published. He succeeded in publishing a paper in *Nature* (July 31, 1966) and a third pamphlet, *The Astronomical Significance of Stonehenge*, which is the only reference on the astronomical potential of Stonehenge now on sale at the monument.

Although Newham died in April 1974, he did live to see some of his interpretations confirmed and sense an increasing appreciation of his efforts by those interested in Stonehenge. As further studies of Stone-henge are completed, it is likely that Newham's place in the history of astronomical investigations there will become more widely known and respected.

How did Newham's results differ from those of Gerald Hawkins? First, Newham reiterated that the only known true astronomical align-ments at Stonehenge included the main axis, or Avenue line, on the summer solstice sunrise, the 92 to 91 alignment on that event, and the 93 to 94 alignment on the winter solstice sunset. Hawkins had equated the axis line with his center-to-Heel Stone line and also had included the reverse directions of the other two alignments.

Newham reminded his readers that the transverse lines 94 to 91 and 92 to 93 were significant, for the Sarsen Circle had been constructed enough off center to permit a view along these lines. If the Sarsen Cir-cle had been exactly concentric with the Aubrey Circle, the view would have been obstructed. From his own survey and analysis Newham pro-posed the following alignments:

ALIGNMENTS	ASTRONOMICAL PHENOMENA
92 to 91	summer solstice sunrise
94 to G	winter solstice sunrise
94 to 93	winter solstice sunset
94 to C	equinox sunrise
92 to G	northern major standstill moonrise
92 to 93	northern major standstill moonset
94 to 91	southern major standstill moonrise

Newham argued that geometrical and astronomical symmetry suggest that a "lost" marker might once have occupied the point he called "G2." If so, good alignments for summer solstice sunset (92 to G2) and for southern major standstill moonset (94 to G2), would have been present. A search for G2 was undertaken, but no such feature was found. The missing sunset line and several lunar alignments were dis-covered later, however, to be indicated by the car-park post holes.

Newham was the first to discover the equinoctial alignment, 94 to C, and to note 94 to Heel Stone could mark the infrequent occurrence of the full moon on an equinox.

Newham and G. Charrière, a French architect, had noted inde-pendently that the astronomical alignments of Stonehenge generate a rectangle. Such a symmetric figure could not have been laid out at lati-tudes very far removed from Stonehenge. A difference of thirty miles, north or south, would transform the Station Rectangle into a par-allelogram. Actually, the two Station Stones and two Station Mounds do not really appear to form an exact rectangle. Atkinson has shown that more displacement of the Station from their present locations than

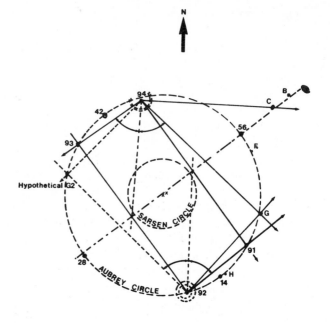

Stonehenge I Alignments (Newham)

C. A. Newham had been quietly investigating the astronomical potential of Stonehenge and discovered before Hawkins a smaller, more reliable set of alignments (compare arrows with Hawkins' Stonehenge I alignments) in Stonehenge I, including an equinoctial line, 94-C. (Griffith Observatory, after C. A. Newham)

can be allowed is needed to obtain true rectangularity. Furthermore, the exact position of the original marker at Station 94 is unknown, for the mound remains unexcavated. Finally, Atkinson has shown that the true latitude for rectangularity would put Stonehenge in the English Channel.

Newham rightly demonstrated that the horizon altitudes as seen from the Station Stones and Mounds permit retention of the figure. Had the horizon profile been different or the ground level altered, the observed risings and settings would have been shifted. This matter of horizon profile is a very important one. Newham observed the actual horizon elevations and as a result discarded many of the reverse alignments that Hawkins retained. Newham considered the magnitude and sense of alignment errors for all possible combinations of lines and found that minimum errors were achieved in any interpretation when observation of first or last gleam was assumed to mark a rising or set-ting. By contrast, Hawkins had assumed that full orb was the observed configuration on the horizon.

Newham's data also failed to support Hoyle's theory of "intentional errors," for no convincing pattern of the sense of the errors emerged for any combination of alignment with first or last gleam or full orb.

Newham reaffirmed that most stones at Stonehenge are really too close together to permit truly precise alignments, but he suggested that more distant outliers may have been used to increase the precision. At the time of Newham's first investigations there was no evidence of dis-tant horizon features like those Thom had found at other sites. As noted above, when an extension to the monument's car park, northwest of Stonehenge, was excavated, three new post holes were discovered about 830 feet from the center of the Sarsen Circle. If the posts that occu-pied these holes were to have been seen from Stonehenge as coinciding with the horizon, they would have had to have been 30 feet high. The post holes themselves have very large diameters, 2½ feet, and excava-tion indicated that wedges were used to keep the posts upright. The large size of the holes and the presence of support wedges both imply that large, tree-size posts were used and may well have been intended to reach the horizon as seen from the monument more than 800 feet to the southeast.

Another series of potential alignments on the car-park post holes was examined by Newham:

ALIGNMENTS	ASTRONOMICAL PHENOMENA
Heel Stone to car park post hole 1	May Day and Lammas sunset
Heel Stone to car park post hole 2	northern minor standstill moonset
91 to car park post hole 1	summer solstice sunset
92 to car park post hole Center	northern major standstill moonset
94 to car park post hole Center	summer solstice sunset
93 to car park post hole Center	northern major standstill moonset

The alignments work in one direction only and in that sense are consis-tent with Newham's Station alignments.

Many interesting and well-thought out ideas about Stonehenge can be found in Newham's pamphlets. He analyzed the Causeway post holes, which fall within a 10-degree arc bounded on one side by the line from the monument's center to the Heel Stone. The arrangement of holes is comprised of six ranks, and Newham suggested that they repre-sent, fairly well, a sequence of yearly observations of the most northerly, or winter solstice, full moon. Over a period of years this full moon would be found to rise farther and farther to the north as the moon approached the time of major standstill. After the maximum northerly

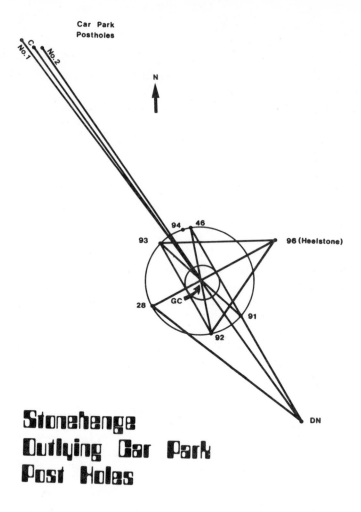

Stonehenge Outlying Car Park Post Holes

The car-park post holes are situated 830 feet northwest of the monument and may have been aligned to the northern moonsets at major and minor standstills. The locations of the post holes have been marked in white on the car-park asphalt. (Griffith Observatory, after C. A. Newham)

position was achieved, the winter full moon would appear slowly to double back, cross the Heel Stone, and reach a maximum displacement to the east side of the Heel Stone 9.3 years later, at the minor standstill. During the next 9.3 years the most northerly full moon would be seen to rise farther and farther north again. In this way an 18.61-year cycle could have been observed. It is possible that this cycle would have been associated with the nineteen-year Metonic cycle of coincidence between date and moon phase.

Three large post holes, in the center of the photograph, discovered during the excavation of an extension to the car park were the first suggestion of long-sightline astronomy at Stonehenge. (E. C. Krupp)

Newham's interpretation of the Causeway post holes, if correct, almost puts us face to face with the people who made them. A permanent marker of the major standstill moonrise is a rather anonymous indication that the moon was watched. It might have been set up during any one of a number of regression cycles in any one of a number of centuries thousands of years ago. By contrast, the Causeway post holes each mark the extreme position of a particular winter's full moonrise. Each row, or rank, of holes is simply a record of the annual extremes during the 4.65 years before the major standstill and the 4.65 years after the standstill. Each row therefore represents half of the 18.61-year cycle. We see here, therefore, the actual record of observations of the northern winter moon. It is almost as if we are looking at a Stonehenge astronomer's data log.

No post holes are known on the other side of the Causeway, and it must be concluded that during this half of the cycle the moonrise extremes were not recorded. The winter full moon, rising once again over the Heel Stone, would have been the signal to start observations again as the standstill approached or to cease observations once the northerly extremes were complete.

Newham compared an actual sequen⊙ of appropriate moonrises with the direction indicated by the Causeway post holes as seen from the center of the Aubrey Circle and found a resemblance between the two. If each of these post holes truly represents a single marker placed by an individual observer and if the six ranks of holes really represent observations made during six consecutive lunar cycles, we have here an actual record of 112 years of moon watching. It is easy to forget that the Stonehengers were real people like ourselves, individuals with personalities. They were members of families. They had weaknesses and skills. We do not often stop to think about the act of a single individual who might leave evidence of his or her presence that lasts thousands of years until it is discovered by us. The monuments and artifacts of prehistory most often conjure up an image of an entire culture, in which the history of individual acts and personal lives is lost. But here at Stonehenge each one of these Causeway post holes was placed in the ground by a specific person who watched a specific moonrise. These subtle post holes, far more than the huge sarsen stones, people the monument once again in our minds with those who built it and used it nearly five thousand years ago, for here we can actually imagine a part of the structure in use.

The Causeway post holes could be remnants of a period of preliminary observation, and the sequence of A post holes may have been a more permanent indicator of the lunar regression cycle. The larger size of the holes suggests a larger, more fixed set of markers. Although it is possible that these A post holes may have contained posts which were used to predict imminent eclipses, it is hard to see how they could have been used to deduce the existence of an eclipse cycle.

Newham speculated that the numbers of certain features of Stonehenge might symbolize aspects of the moon's behavior. The thirty Sarsen Circle stones, for example, might each represent one day of the synodic month, the period from, say, full moon to full moon. This period of time is actually closer to 29½ days, but perhaps, offered Newham, the smaller size of stone 11 intentionally symbolized the half-day. A quick multiplication shows that fifty-nine days are included in two synodic months. The fifty-nine Y and Z holes may have been intended to represent this period of time. It is uncertain how many bluestones actually were included in the Bluestone Circle, but estimates of fifty-nine, sixty, and sixty-one have been made. The first estimate takes on added significance if lunar cycles are considered. The Bluestone

Stonehenge
Causeway Post Holes

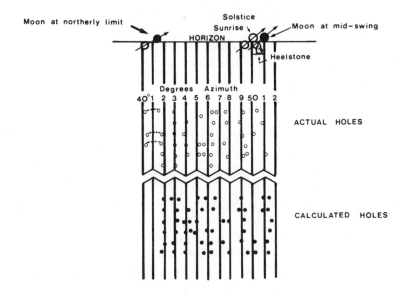

The actual positions of the Causeway post holes are shown by open circles upon a set of lines that represent the direction, in degrees of azimuth, east of true north. The direction of the most northern moonrise and the direction of the summer solstice sunrise, as well as the position of the Heel Stone and the position of the northern moonrise midway between standstills, are all shown as they would be seen from the center of Stonehenge. The actual post holes fall in the zone between the moon at midswing and the moon at the northernmost extreme. Newham compared these with a set of calculated holes, shown below, which would have indicated the yearly northernmost moonrises through a period of time when Stonehenge was in use. Newham showed that the pattern of actual holes is similar to a typical pattern expected from a century or so (six consecutive standstill cycles) of real moonrise observations. (Griffith Observatory, after C. A. Newham)

Horseshoe, with its nineteen stones, could have represented the nineteen years of the Metonic cycle, but here, again, caution is advisable in any discussion of the ruined bluestone features.

Perhaps the most intriguing Stonehenge number of all is 56, for the fifty-six Aubrey Holes. Prior to Hawkins no one had offered any compelling interpretation of the number and nature of the Aubrey Holes. Hawkins' clever use of the Aubrey Circle must have contributed to the over-all impact of his Stonehenge theories. Doubtless many a reader of *Stonehenge Decoded,* upon learning of Hawkins' explanation of the enigmatic fifty-six holes, must have felt convinced that at last, if simply for want of any other solution, the mystery of Stonehenge was nearly revealed.

Hoyle, although dissatisfied with the Hawkins eclipse predictor, was sufficiently intrigued by the idea of a connection between the 18.61 year regression of the nodes and the fifty-six Aubrey Holes to devise an eclipse predictor of his own. The Hoyle eclipse predictor, as already shown, was an improvement over the Hawkins model and in principle predicted all eclipses. We have also seen, however, that calibration of either of these eclipse predictors would have been extremely difficult and would have required foreknowledge of cycles of eclipses.

So much emphasis had been placed upon the Aubrey Circle and eclipse prediction that Newham felt justified to propose a series of alternate explanations of the fifty-six holes. He personally favored one of these because he felt that a circle divided into fifty-six parts was particularly well-suited to investigation of lunar phenomena. The average spacing between two Aubrey Holes is 6.428 degrees; four holes span, on the average, 19.286 degrees. The angular difference between the azimuths of major and minor standstills at Stonehenge is 19.26 degrees. Newham therefore suggested that the layout of the fifty-six holes might have been based upon the moon's angular motion from major to minor standstill. The positions of all fifty-six holes might have been determined by marking points on the Aubrey Circle's circumference at the correct arc length for the 19.26 degree angle. Once around the circle would generate the first eighteen positions. The arc length "ruler" would then overlap the position of the initial hole by about one third of the arc length. This spot would mark the next sequence of holes, and finally, by continuing around the circle again, the rest of the Aubrey Hole positions would be given.

To Newham, it appeared that the attention of the Stonehengers was concentrated on the behavior of the winter moon. Observations of the winter full moon may have been systematically recorded by posts that once occupied the Causeway, and against this sequence of observations the 18.61-year regression cycle could have been established. The bright

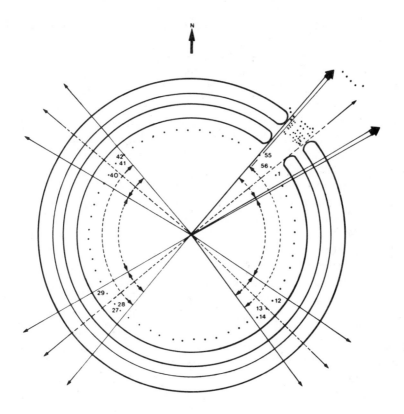

Stonehenge Lunar Alignments (Newham)

Newham was also interested in the placement of the Aubrey Holes. In this plan of the outer portions of Stonehenge, the Aubrey Holes are shown as a circle of dots inside the ditch and bank. In the opening on the northeast side of the bank, a large number of dots represent the Causeway post holes, and beyond these, farther northeast, are dots representing the A post holes and the Heel Stone. Arrows to the northeast, northwest, southeast, and southwest pass by certain Aubrey Holes, which are numbered. These arrows mark the directions of the lunar standstill moonrises and moonsets. The arrows intersect the Aubrey Holes on the north side, but they miss the mark on the south side. Newham speculates that the Stonehengers set out the Aubrey Holes to coincide with the standstill directions, but, he adds, because they did not know about the effect of parallax on the apparent directions of moonrise and moonset, their plan failed on the south side of the Aubrey Circle. Perhaps in jest, Newham suggested that this is why the Stonehengers filled up the Aubrey Holes so soon after digging them. They abandoned their folly. (Griffith Observatory, after C. A. Newham) winter full moon sails high through a long and otherwise dark winter night, and its practical benefit may have also prompted interest in its behavior.

If the ancient Stonehenge astronomers spaced and positioned the Aubrey Holes with respect to winter standstills, the risings and settings at major and minor standstills would have been marked by Aubrey Holes 39, 42, 55, and 2, and, in fact, these Aubrey Holes do mark the boundaries of the winter moon. Newham imagined that lines for the winter moon, from the center of the Aubrey Circle to the Aubrey Holes already mentioned, might have also been extended to the southwest and to the southeast, under the assumption that the less well-studied summer moon would rise and set symmetrically with the winter moon. Although moonrise and moonset extrema for summer major and minor standstills are approximately opposite the winter positions, the symmetry is not exact. The moon's orbit is elliptical, and the moon is sufficiently close to the earth to permit the difference in distance to affect the alignments. Newham pictured the Stonehenge astronomers as careful observers of the winter moon, who, after laying out the Aubrey Holes and predicting the correct positions for the summer moon abandoned the project and filled up the holes when they discovered that the summer moon didn't behave as expected. Geocentric parallax was the culprit, but there was no way they could have known that. Modern research suggests that the summer moon was studied at Stonehenge, but we shall probably never know if the research was in response to the failure of an Aubrey Circle experiment. Perhaps Newham's explanation of the Aubrey Holes is too pat, but at least it affords a very human explanation for an otherwise enigmatic feature of Stonehenge.

A WELL-HEELED MOONRISE

Until Newham, no self-respecting interpreter of Stonehenge astronomy could neglect the monument's solstitial alignments. For many, the view of the summer solstice sun, weather permitting, climbing above the northeastern horizon and standing poised for a moment upon the pointed tip of the Heel Stone's silhouette, was the most visually convincing evidence of Stonehenge's astronomical potential. Motion picture footage of this event was incorporated in the 1965 CBS television production *Mystery of Stonehenge*. Throughout the program the claims of Hawkins were debated until, at the conclusion, a moment of high drama was achieved when the theory was tested, and seemingly confirmed, by the image of the sun's dancing disc above the Heel Stone.

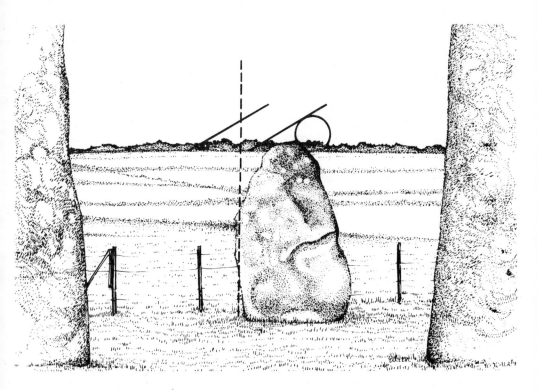

The Heel Stone is touted as a good summer solstice sunrise indicator, and the slanted line to the left of and touching the edge of the sun represents the path of the upper limb of the sun as it rises on the summer solstice today. It is clear that the sun does perch for a moment full orb above the Heel Stone at the solstice. The sunrise hasn't always followed this path, however, and the second slanted line, to the far left, represents the path of the upper limb of the rising sun in 1500 B.C. Clearly, by the time the sun reached the Heel Stone it was well above it and not poised at the Heel Stone's tip. The vertical dashed line represents the true axis of Stonehenge, and it is obvious that the Heel Stone is offset from this axis. (Griffith Observatory, after C. A. Newham)

It is true that today the Heel Stone is a good indicator of the direction of the summer solstice sun for the full disc tangent to the northeastern horizon. The Heel Stone, however, does not mark, nor has it ever marked, the first gleam of the solstitial sunrise. Even though the Heel Stone provides an excellent stage for latter-day Druids, it is only in the present era that the Heel Stone successfully has marked the full orb of the summer solstice sunrise.

In previous centuries, as now, the first gleam of the sunrise appeared to the left of the Heel Stone as viewed from the center of the Sarsen

Circle. After first gleam, the sun rises at an angle to the horizon. Almost ½ degree, or one solar diameter, separates the position of first gleam from the tip of the Heel Stone.

The sun's positions of first gleam and full orb at Stonehenge are determined by the angle of the obliquity of the ecliptic. Although this angle changes only very slowly, the rate is sufficient to alter the point of sunrise over several centuries. In prehistoric times the change was roughly 40 arc seconds per century.

In 1500 B.C. the position of first gleam was even further to the left of today's position and about 1½ degrees north of the Heel Stone. By the time the sun rose directly above the Heel Stone, its lower limb would have been a full solar diameter above the horizon. Even when it was thought that the astronomical alignments dated from 1500 B.C., the Heel Stone had to be regarded as a dubious solstitial marker. The recalibrated radiocarbon dating scheme has now pushed Stonehenge I, and the Heel Stone, back to 2800 B.C., approximately. The point of first gleam is accordingly shifted still farther to the left, above five solar diameters. When the sun finally positioned itself above the Heel Stone, its lower edge cleared the horizon by a few solar diameters.

Discrepancy in the alignment of the Heel Stone with the summer solstice sunrise should really come as no surprise to us. Early investigators of Stonehenge astronomy usually claimed that the monument's axis, or the Avenue, was aligned on the solstice. The Heel Stone is not centered in the Avenue but is located slightly right of center. For Hawkins, this offset was more a reflection of the limited skill of the builders than a clue to some other purpose. Doubtless the now-slanted Heel Stone once stood upright and therefore reached higher to touch the lower limb of the sun of 1500 B.C. Or so Hawkins argued. Atkinson pointed out that an upright Heel Stone would also have its tip moved farther to the right and away from the correct sunset line. Atkinson also disputed that the tip of the Heel Stone ever protruded above the horizon and doubted that Hawkins' Heel Stone solstitial alignment was valid. Atkinson did concede that the Heel Stone may have served as a valuable eclipse signal when the winter solstice full moon rose above it.

It may seem particularly frustrating that such a strong tradition as the Heel Stone sunrise is without foundation, but a variety of escapes are available. We don't know, for instance, at what eye level observations were intended to be made. Perhaps the Heel Stone was deliberately offset, in the manner of Hoyle's intentional "errors," to permit interpolation of the solstice date. Or possibly, by Stonehenge IIIa, when the Slaughter Stone and its missing mate were erected on the axis, they were intended to frame the Heel Stone and provide a window in which the sun would ceremoniously appear at an arbitrary height above the tip of the Heel Stone. We might even imagine that the approximately solstitial alignment of the Heel Stone prompted subsequent inhabitants of the Salisbury Plain to conclude, erroneously, that the solstice was the intended target.

An entirely different interpretation of the Heel Stone was stimulated by Newham's emphasis on the lunar aspects of Stonehenge. Jack H. Robinson argued in a letter in *Nature* (1970) that the Heel Stone's slight offset from the Avenue's center line made the Heel Stone an ideal moon indicator.

Hawkins had also noted that the full moon rising over the Heel Stone at the winter solstice could have signaled an imminent eclipse, but he maintained that the Heel Stone was first and foremost a solstice sunrise pointer. He viewed the Avenue as an averaged "best fit" for both the summer solstice sunrise and the winter solstice sunset. The Heel Stone was offset because it was intended to mark precisely only the former.

Robinson agreed that a winter Heel Stone full moonrise could have announced an eclipse, but he also maintained that this was the primary and exclusive function of the Heel Stone. The offset resulted from the relative nearness of the moon in comparison with the sun's distance from the earth. A lunar eclipse could occur only when the moon was exactly opposite the sun in the sky, and near the time of winter solstice the moon must therefore occupy the symmetrically located summer solstice position of the sun. Because the moon is so much nearer than the sun, its geocentric parallax shifts it nearly a degree below the horizon relative to the summer solstice sunrise position. As a result, when the moon finally rises, it rises, 1½ degrees to the right of the solstitial line and directly above the offset Heel Stone.

PROFESSOR THOM AT STONEHENGE

It should be evident by now that a credible interpretation of Stonehenge depends critically upon accurate on-site survey data. Newham's emphasis on the lunar significance of Stonehenge is supported by the painstaking on-site information he himself obtained. The increasing interest in Stonehenge astronomy, stimulated by Hawkins, obviously justified a comprehensive, reliable survey of the entire monument. Just such a survey was undertaken by Professor Alexander Thom and his associates in 1973. All features showing on the surface were carefully surveyed, and the best information on the positions of features known only by excavation were incorporated as accurately as possible into the final plan.

Once Thom had completed the survey, he examined Stonehenge for the kind of Megalithic geometry, metrology, and astronomy he had extracted from so many other sites, and he demonstrated that at least some aspects of Stonehenge were consistent with the same principles of design he had found elsewhere.

Atkinson had noticed that the inner faces of the Sarsen Circle uprights and of the trilithon uprights had been worked flat and smoothed. The outer sides, by contrast, were rough and unfinished. Thom based his geometric analysis on the inner faces and found that the inside circumference of the Sarsen Circle is 45 MR (or 112.5 MY). Each Sarsen Circle upright is therefore permitted a width of 1 MR, and the spacing between uprights is ½ MR. According to Thom's previous work, Megalithic people preferred to lay out straight lines in whole numbers of Megalithic yards (1 MY=2.72 feet) and perimeters in integral multiples of Megalithic rods (1 MR=2.5 MY=6.803 feet).

Four of the five trilithons appear to be contained between two Megalithic ellipses. Such ellipses are constructed with major and minor axes and the distance between foci all integral in Megalithic yards. Only the inner ellipse completely fulfills these conditions. The major axis is 27 MY, and the minor axis is 17 MY. The distance between foci is 21 MY. Major and minor axes of the outer ellipse are 30 MY and 20 MY respectively. The trilithon uprights appear to have been measured at 1 MR. Each trilithon archway is ¼ MR wide.

The Aubrey Circle is set out with a circumference of 131 MR. The Y and Z holes are located on open rings, each of which is constructed from two semicircles. The radii of the Y hole semicircles are 12½ and 13 MR, and the radii of the Z hole semicircles are 9 and 9½ MR. Both Y and Z holes lie on radii that pass through the Sarsen Circle upright centers. The Bluestone Circles are in ruins and have had a complex history, but what remains of them does not seem to fit a Megalithic geometry.

STONEHENGE: THE NEW ASTRONOMY

Professor Thom's examination of other Megalithic sites provided many examples of precision solar and lunar observatories. Typically, such a site would include a backsight, marked by standing stones, and a distant foresight, perhaps a notch or a peak on the horizon. Often the lunar observatories also included ancillary distances or grids marked on the ground that could have been used to interpolate the desired position of the moon. At Stonehenge a search for distant foresights was inaugurated, and the earlier discovery of the car-park post holes encouraged the search.

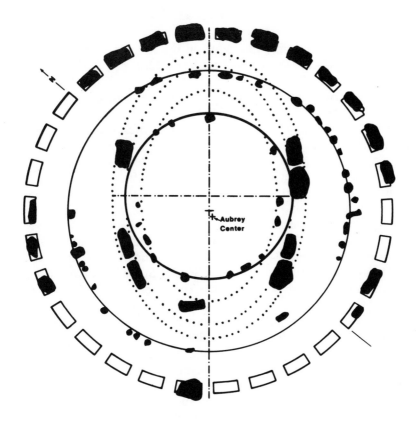

Thom's interpretation of Stonehenge geometry encloses the Sarsen Circle stones within two circles, 48 and 45 Megalithic rods in circumference. (Griffith Observatory, after Alexander Thom)

A new wrinkle was added to old summer solstice sunrise controversy when Newham pointed out a small mound to the Thoms. The mound, which has been called Peter's Mound in memory of Newham (known to his friends as Peter), is on the axis of Stonehenge, 8,981 feet northeast of the center of the Aubrey Circle. Its position gives a good azimuth for the center of the sun's disc at 2700 B.C. Of course, it is not possible to be sure that the center of the sun's disc was intended, nor for that matter is it known if the mound is prehistoric. It is sufficiently close to the solstitial alignment to warrant further investigation, however, and it may have been intended as the stage upon which the rising sun made its solstitial appearance.

Lunar lines and possible foresights on the horizon were also hunted. Of the eight possible lunar standstill lines, four, or possibly five, may have been found.

The northern major standstill moonset line extends from Station Mound 92 to Station Stone 93 northwest across the car-park post holes and across Fargo Plantation to Gibbet Knoll, near Market Lavington and 9.16 miles from Stonehenge. Gibbet Knoll is a curious, small earthwork. What remains is little more than a long ledge about 4 feet high. Although it is not known if Gibbet Knoll dates from Megalithic times, it is in exactly the right spot for the lunar foresight. Something larger than the present mound would be needed to observe a foresight silhouetted against the moon, but Thom estimates a pile of brush, about 15 feet wide and situated on a mound of earth, would do the job.

Another lunar alignment extends to the southeast. By starting at Station Mound 94 and continuing across Station Stone 91 and on to a hole of uncertain origin 617 feet from the center of Stonehenge, an alignment is provided which intersects the Iron Age hill fort Figsbury Ring 6.6 miles away. Figsbury Ring is a much more recent structure than Stonehenge, but of course it is possible it was used as a foresight in Megalithic times. There is a low mound in the center of the fort's central flat area, and its position is reasonably close to the southern major standstill moonrise.

Hanging Langford Camp, 8.04 miles to the southwest of Stonehenge, provides a convenient ridge for a foresight for the southern minor standstill moonset. No Megalithic features are apparent there, however. Similarly, Chain Hill, 3.7 miles northwest of Stonehenge, is an ideal location for the southern major standstill moonset foresight. Finally, a tumulus remains visible on Coneybury Hill, 5,000 feet to the southeast of Stonehenge. The feature could have been used to observe the southern minor standstill moonrise.

The other three possible lunar lines remain uninvestigated, and, in fact, winter standstill moonrise foresights will be particularly hard to locate, for the land in that direction is considerably built up with roads, houses, and gardens.

In a sense, Thom's version of Stonehenge astronomy is Carnac astronomy turned inside out. At Carnac, in Brittany, Thom concluded that Le Grand Menhir Brisé, the monstrous sixty-eight-foot menhir whose top probably stood sixty feet above the ground, was used as a universal foresight for several backsight stations in the general Carnac area. Stonehenge appears to have been a universal backsight for a series of horizon foresights.

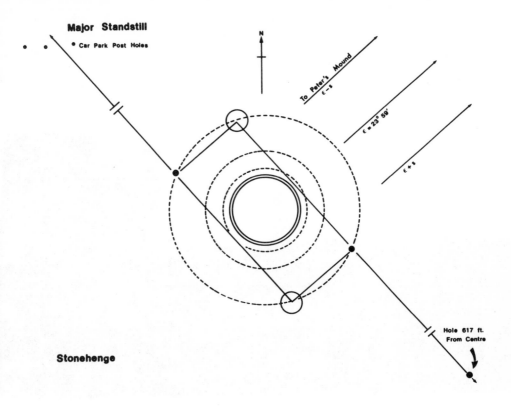

The Station Mounds and Stones nearly form a rectangle. The long sides of this figure extend to distant foresights along lunar alignments. (Griffith Observatory, after Alexander Thom)

Still more astronomical interpretations of Stonehenge have recently emerged. Richard Brinckerhoff, a science teacher in New Hampshire, has examined cupmarks on top of some of the Sarsen Circle lintels still in place and has concluded that some of them were used with auxiliary equipment to measure the position of the sun at the summer solstice sunrise and the position of the moon at the northern major standstill moonrise. A New Zealander, A. D. Beach, has interpreted the Aubrey Holes as tidal amplitude predictors for seagoing Stonehengers. Atkinson has commented skeptically on both of these ideas. Finally, Fred Hoyle, in a new book, *On Stonehenge*, has expanded on his original ideas.

The view from Station Mound 92 to Station Stone 93 indicates the northern major standstill moonset, to the northwest. Station Stone 93 is the small stone beyond and to the left of the outermost Sarsen Circle upright. The trees on the horizon are Fargo Plantation, and Gibbet Knoll is several miles beyond. (E. C. Krupp)

Small trees now grow along a low bank, about 4 feet high, on Gibbet Knoll, and the bank may have marked the location of a distant foresight. Alexander Thom estimates that a 15-foot-high structure, built on this spot could have been visible against the moon's disc from Stonehenge. (E. C. Krupp)

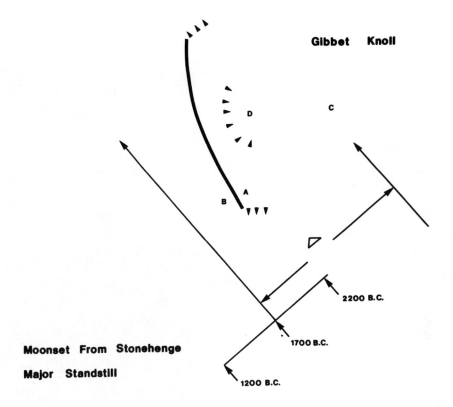

Arrows indicate the direction of the northern major standstill moonset alignment, which extends 9.16 miles northwest of Stonehenge to Gibbet Knoll, near the village of Market Lavington. Points A, B, C, and D mark different ground levels around the bank, which is highest at A and along the solid curve. If the earthwork there actually was used, it may date from the later periods of Stonehenge, which is in the direction opposite the arrows. (Griffith Observatory, after Alexander Thom)

By this stage we need no longer doubt whether Stonehenge had astronomical significance. Instead, we might marvel that it had so much significance. Its builders and users were capable of remarkable achievements through their own tenacity and their strongly practical and observational approach. It only remains for us to equal their genius and deduce what the deuce they did.

A line of sight southeast from Station Mound 94 to Station Stone 91 marks the southern major standstill moonrise. This line of sight skirts just left of the Sarsen Circle. The trees in the distance block the view of the horizon. The arrow points down to Station Stone 91, near which a person can be seen standing. (E. C. Krupp)

Figsbury Ring

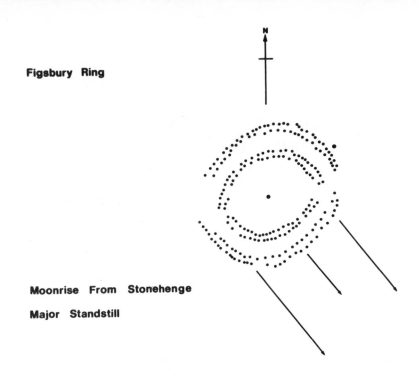

Moonrise From Stonehenge

Major Standstill

The southern major standstill moonrise alignment intersects Figsbury Ring 6.6 miles from Stonehenge. The contours of Figsbury Ring are marked by dots and the arrows indicate the moonrise direction. Stonehenge is in the direction opposite the arrows. (Griffith Observatory, after Alexander Thom)

Figsbury Ring is an Iron Age hill fort. Its earth ramparts can be seen at the crest of the hill in this view, and an entrance gap is visible at the center of the profile. (E. C. Krupp)

Many unorthodox interpretations of the original layout of Stonehenge and the meaning of the number 56 have been put forth. Here, sarsen arches prevail. (Stockdale and Veze)

CHAPTER FOUR

Archaeoastronomy of North America: Cliffs, Mounds, and Medicine Wheels

JOHN A. EDDY

Although we have suspected there is something astronomical about Stonehenge for over two hundred years, the ruins seemed unique until Alexander Thom revealed the geometrical and astronomical complexity of the many other Megalithic sites. It comes now as an even greater surprise to us in the United States that not all ancient astronomical sites are in distant lands. Some are as close as the American Midwest and the Rocky Mountains. Dr. John A. Eddy, senior staff member at the High Altitude Observatory in Boulder, Colorado, tells here, for the first time, the whole story of archaeoastronomy in North America.

We are much closer in time to the cliff dwellers of the Southwest, the mound builders of the Mississippi Valley, and the nomadic peoples of the Great Plains than we are to the builders of Stonehenge, but, as John Eddy shows, we encounter the same major problem as we did with the Megalith Builders: the American Indians left us no written record to tell us what kind of astronomy they did. It is necessary, therefore, to rely on careful surveys of remaining sites and structures. Without these there would be no evidence at all.

Eddy is particularly well qualified to evaluate the North American evidence. He is a solar physicist, but he has also authored papers on a variety of historical aspects of astronomy, including Thomas Edison's contribution to infrared astronomy and, more recently, the Maunder minimum in solar activity. Most importantly, Eddy surveyed the Big Horn Medicine Wheel in Wyoming and demonstrated its astronomical importance in a paper which was featured on the cover of Science. *Discovery of this "American Stonehenge," as news reports dubbed it, has re-emphasized how important a role astronomy may play in many ancient cultures.*

Eddy's first-rate work on the Big Horn Medicine Wheel maintained the high standards already set by Professor Thom in Europe. Several other medicine wheels and related structures have since been surveyed by Eddy, and they agree with his original interpretation. He is especially aware of the pitfalls of archaeoastronomical research and reports in this chapter what is and isn't yet known about the astronomy of the North American Indians.

THE INDIAN AND THE SKY

It is said that when the first Europeans came to the North American continent and in time were able to converse with Algonkian tribes, they pointed to the familiar pattern of seven stars in the northern sky and said, "Those stars are what we call 'the Bear.'" The Indians, so the story goes, replied, "Yes, we call them 'the Bear' too."

It is hard for anyone to find a bear or any part of a bear in the seven stars of the Big Dipper; the European basis for calling them a bear, Ursa Major, is lost in dim origins of prehistory. And so the Indians' reply poses an interesting riddle: why would two cultures which had developed separately in isolated halves of the world give the same unlikely name to a pattern of seven stars in the northern sky?

A romantic and intriguing answer is that in this innocent exchange European and Indian revealed their common ancestry, that the unlikely association of this star pattern with a bear had its origin in the murky thoughts and traditions of primitive people on a distant continent, and that these traditions were preserved and carried in opposite directions by diverging peoples over tens of thousands of generations, westward into Europe and eastward to Asia, up across the Bering Strait and down, to meet again in chance conversation on the shores of newly found America.

But an answer equally good is that the name was chosen independently for natural and logical reasons: in the northern hemisphere the north is cold and bears are hardy animals. The circling of the constellation around the celestial pole resembles the restless prowling of a bear around his den. Alternatively, of course, the choice of the same name by isolated peoples could have been a coincidence of pure chance.

Whatever the answer, and whether or not the story related at the beginning of this chapter is true at all, by entertaining the question we accept the belief that the American Indians were interested in the sky. I do not think that many would doubt this or question that an interest and knowledge of things astronomical were practical and important parts of North American Indian life. We find the stars and the sky woven through their legends, their symbols, and the basic fabric of their practical religion, in which the sun was principal deity.

In admitting that the Indians measured the passage of time in "moons," we acknowledge that they were practiced observers of lunar phases. Since they spent more time than we in looking at the night sky (and under far better conditions), we may assume that they knew the star patterns better than most of us do today. We may equally well assume that a part of their natural lore was a familiarity with the changing path of the sun against the star background and that an important part of this practical knowledge was an awareness of the summer and winter turning points of the sun's place of rise and set. It is natural that the Indians would associate the sun's approach to the one turning point, in summer, with light and warmth and a bountiful earth, and the approach to the other, in winter, with cold and darkness and barren land. We know that they designed rituals to celebrate the solstices— the one a time of rejoicing and the other a time to strengthen the weakened sun for its long journey northward. These truths are self-evident. They follow logically and naturally in the lives of any people who live by the sun, out of doors, whose whole existence is dominated by the seasons, and who watch for dawn from rude beds on the hard ground. We hardly need any archaeological evidence to establish whether or not the American Indian had a practical astronomy and an intimate knowledge of the sky, although, in fact, there are abundant signs scattered over the continent in the form of rock paintings and rock carvings, in celestially aligned mounds and other structural relics, and in stone patterns laid out on hills and mountain tops.

Curiously, we do not read much of the Indians' astronomy in the first-hand accounts of their existence made by observers in the final centuries of the Indians' natural life on the continent. We find tales and legends of the sky in a mythological sense but descriptions of practical use of the sky are virtually absent from the depositions taken from Indian descendants in this century and the last. We do know that the Pueblo Indians designated tribal officers to identify the times of solstice. They did this into modern times by setting up horizon markers for sunrise and sunset. But this is almost the entirety of the first-hand, direct evidence of the North American Indians' practical interest in the sky. The Plains Indians did not know, or admit knowing, either the origin or the obvious astronomical alignment of the rock patterns that are found in their lands. Those who were still using the Mississippian mounds when Europeans came told or knew little about the details of their construction and plan.

The absence of astronomical reference in depositions and first-hand accounts is more than anything an indication of the limitations of this technique in ethnology. In first-hand observation and depositions, we take a real sample of only the last 1 or 2 per cent of the time of residence of the Indian in America and make further deductions only by hearsay and supposition. Moreover, in the sampling process we may influence the answers. It is much like the uncertainty principle of quantum mechanics: the act of making a measurement influences the outcome. In the anthropological application, the act of inquiring affects the answers given; moreover, the time required to learn to communicate with the Indians allowed their ways to shift and change. They were quick to adopt the ways of the white man, especially as they appeared better or easier than the Indians'. The bow was readily dropped for the rifle, and earthen cooking pots were abandoned when metal kettles could be had. In but a generation or two natural skills and practices, such as telling time and season by the stars, could be lost and forgotten through replacement by easier, newer ways. I cannot harness a workhorse nor hitch a team to a wagon, yet to my grandfather these were acts of second nature. So in a generation or two at a time of revolutionary change in Indian life, much practical lore could have been lost.

Depositions are further suspect, for the Indians were often inclined to give the answers sought—to create the mythical seven lost Cities of Cíbola of the American Southwest and direct the seekers there—if that was what was asked. In this trait we find another possible answer to the riddle of the Ursa Major stars: acculturation by emulation. The Indian called them "the Bear" because the white man had.

Unfortunately we cannot get to the truths of the North American Indians' use of astronomy through written evidence, for they had no written language. Yet the many marks and lines left on cliffs and walls and hilltops are an objective language of deeds, which can be read as evidence that is probably more eloquent and compelling than spoken words or repeated tales, however sincerely told.

ASTRONOMICAL RECORDS IN INDIAN ROCK ART

A. Possible Records of the Crab Supernova

If we search the sky near the tips of the horns of the celestial bull, Taurus, we will find, with the aid of a telescope, a nebula of great beauty called "the Crab." It is an object of prime astrophysical interest, for it is believed to be the remnant of a catastrophic explosion of a star, a supernova outburst. From present-day measurements of the expansion rate of the nebula, astronomers calculate that the supernova explosion should have been observed in Taurus about nine hundred years ago, in the middle of the eleventh century, about the time of the First Crusade in Europe. The supernova would have appeared as a new star of extraordinary brilliance.

In fact, Chinese records of the Sung dynasty indicate that a bright object was observed in Taurus in A.D. 1054. It was recorded in the Chinese annals as a new star, or "guest star," and was first seen on July 5 or 6. It was bright enough to be seen in the daytime for about three weeks, and it then slowly faded in brilliance. At its first appearance it was probably the brightest object, other than the sun and moon, ever seen in the sky in the record of humanity. At its brightest, it was about five times brighter than any of the planets we see in the night sky today.

Other peoples on the earth should have seen the bright new star and recognized it as something unusual, if they were regular watchers of the sky. If the American Indians saw and recorded it in 1054, it would indicate that they knew the sky well enough to recognize a new feature in it.

In 1955 two unusual Indian pictographs were found in northern Arizona, in the course of archaeological fieldwork. Though they were found many miles apart, the two drawings showed a similar design: a crescent moon near a solid circle. One, a painted pictograph, was found on the wall of a cave. The other, a petroglyph (carved in stone), was on a canyon wall. They were considered unusual because the crescent was not believed to be common in Southwest Indian rock art.

William C. Miller, a distinguished astronomical photographer at Mt. Wilson/Palomar (now Hale) Observatories, became interested in the drawings, on the supposition that they might represent a real astronomical event. Might they portray Amerindian observations of the brightest object, other than the sun and moon, ever seen in the sky, the supernova explosion in Taurus of A.D. 1054?

To help answer the question, Miller needed to know whether or not the bright new star seen by the Chinese was near the moon in the sky at the time of its discovery. If it was, if the moon was in crescent phase at the time, and if the sites were occupied in A.D. 1054, then the association would seem plausible. These three questions were not hard for astronomy and archaeology to answer. Calculations showed that the supernova should have appeared very near the moon on the morning of July 5, 1054, as seen from western North America and that the moon was in crescent phase. Moreover, the moon appeared above the star in the sky, as shown in both of the Arizona drawings. Archaeologists verified that the two sites were indeed occupied at that time, in the middle of the eleventh century. The case, albeit circumstantial, seemed strong that in at least two neighboring locations in the American Southwest, Indians saw the bright new star near the moon and recorded it on the walls of canyon and cave. If true, it would indicate that these early Americans knew the sky well enough to recognize a newcomer in the field of stars and that, in a limited sense, they knew the sky as well as the more culturally advanced Chinese astronomers.

This is an interesting and important point, and since the time that Miller published his finds a number of other American scientists have pursued the problem further, carrying out more details of the calculations of star and moon positions and searching for other examples of the possible portrayal of the event by early American Indians. These scientists, all astronomers, who have pooled their work on the problem, are John C. Brandt and Stephen Maran at the NASA Goddard Space Flight Center, Greenbelt, Maryland; Ray Williamson of St. John's College, Annapolis, Maryland; Robert Harrington of the U. S. Naval Observatory, Washington, D.C.; and Von Del Chamberlain of the National Air and Space Museum, Washington, D.C. Others who worked on the problem were Muriel and William J. Kennedy of Pacific Grove, California, and Ranger Clarion Cochran of the Chaco Canyon and Aztec Ruins National Monuments in New Mexico. Together they have found and examined a significant number of additional examples of unusual pictographs and petroglyphs which portray a conjunction of a crescent moon with a starlike object. Each of these depictions is found at a site which in all probability was occupied at the time of the supernova appearance, and each is in the vicinity of a clear view of the eastern sky, where the moon and star appeared near the horizon, before dawn on July 5, 1054. Brandt and his colleagues have determined that the conjunction between the crescent moon and the supernova was closest for observers in the western part of the North American continent, which adds to its importance there, strengthening the case that

The crescent moon was visible in western North America near the Crab supernova in the predawn eastern sky on July 5, A.D. 1054. (John A. Eddy)

Several examples of crescents in rock art from the American Southwest are associated with what may be representations of the A.D. 1054 supernova. (John A. Eddy)

below: Pictographs & Petroglyphs from sites occupied in the 11th Century.

Chaco Canyon, New Mexico

Northern Arizona

Northern Arizona

Fern Cave California

Symbol Bridge California

Abo Monument New Mexico

the spectacle might have been remembered and recorded. By the time that the moon rose in China on the early morning of the conjunction, the separation between moon and star had increased to about 6 degrees—three times the distance when it appeared to the inhabitants of America.

In attempting to decipher astronomical signs and messages in the Indian pictographs and petroglyphs, astronomers and archaeologists are employing techniques similiar to those used by Alexander Marshack on the drawings and relics of Ice Age peoples in Europe. In each case we deal with a partially translated language of symbols. In each case there are genuine and serious doubts whether the interpretation is accurate in detail or in general. And in each case reliance, in the end, must be put on the number of cases that are found which seem to fit a certain interpretation. Doubts will always remain. The numerous added examples of possible pictures of the 1054 supernova have strengthened the original case made by Miller, but Brandt and his colleagues are quick to point out that they can never prove that the rock art records refer, in fact, to that particular object in the sky.

Professor Emeritus Florence Ellis of the Anthropology Department of the University of New Mexico is one who doubts the association of these markings with the supernova event. From her knowledge of present customs of the Pueblo Indians in the area, it seems unlikely to her that any unusual sky event would have been recorded, even if it had been seen and recognized as something special. Moreover, she feels that the crescent symbol was not uncommon and that it was used to represent the sun, Venus, or the new moon in connection with the Indians' way of counting time by lunar phases. Professor Ellis feels that the crescent and star symbols probably identified sites from which sunrise was watched to mark the time of solstice. Sun symbols are found near some of the crescent and star markings which have been cited by Brandt and his colleagues.

B.   Astronomical Interpretation of Other Symbols

Von Del Chamberlain, one of the investigators of the possible depiction of the 1054 event, has pointed out in other work that we are still ignorant of what symbols the Indian may have used for many possible astronomical objects. A rayed-disk symbol for the sun is common in Indian pictography and is well recognized. But how, for example, might the eclipsed sun have been shown? If the Indians recorded the bright supernova, they would likely have recorded the unusual spectacle of a total solar eclipse. Such an event might well have worked its way into lore, ceremony, and rock art depictions. If such a record could be found, it would provide the possibility of rather precise dating. As evidence that eclipses were noted and recorded, at least in modern times, Chamberlain cites a possible depiction of a solar eclipse in a pictographic diary, or "winter count," of the Dakota tribe for the year 1869–70. A total solar eclipse was visible in their lands on August 7, 1869. Chamberlain suggests that in rock art some pairs of circles, one circumscribed in the other, may be the representation of the eclipsed sun.

Stars and star patterns were recorded by Indians in a number of ways which have been described by Chamberlain. The most common, perhaps, is a four-pointed star or cross, sometimes with circumscribed circle. Stars might also be shown as small circles and as dots and rayed points. There are many known examples of patterns representing stars and perhaps even constellations carved, pecked, or painted on rocks and on the ceilings of caves and other structures. Stars, asterisms, and constellations appear often in Navajo sand paintings.

Claude Britt, Jr., of Round Rock Trading Post, Chinle, Arizona, has recently described a great number of Navajo "star maps," or star ceilings, which are found in the Canyon de Chelly ruins in Arizona.

Chamberlain has also cited the Indian's interest in meteors, fireballs, and meteorites, which confirms that the Indian observed and thought about "falling stars." A legend of the Menomini Indians of the Great Lakes area, cited by Chamberlain, reveals that an association was made between "shooting stars" and iron meteorite fragments found on the ground:

> When a star falls from the sky
> It leaves a fiery trail.
> It does not die.
> Its shade goes back to its own place to shine again.
> The Indians sometimes find the small stars
> where they have fallen in the grass.

He points out that a number of iron and stony meteorites have been found in Indian ruins and burial mounds, in places and circumstances which indicate that they were considered to be of especial importance or worth.

ASTRONOMICAL ORIENTATIONS OF ANASAZI STRUCTURES

The Anasazi were prehistoric basket makers and cliff dwellers whose civilization thrived in the Four Corners region of the American Southwest beginning nearly two thousand years ago. They are known to most of us as the builders of elaborate celled communal dwellings, underground circular kivas, and cliff houses. They are the same people who were mentioned in the preceding section as having possibly depicted the supernova explosion of 1054. The Anasazi were the ancestors of the present-day Pueblo Indians of New Mexico and Arizona, some of whom, as at the Taos Pueblo, still live in similar structures and follow a similar way of life.

Some modern Pueblo Indians continue to use the rising and setting points of the sun at solstice to mark their year and its festivals. We may suppose, therefore, that their ancestors put astronomy to use in much the same way, and we would expect to find evidence of solstitial alignments and possibly other celestial alignments in some features of the Anasazi architecture. In addition, the regular lines, angles, and occasional symmetry of some of the ruins suggest that some form of rudimentary surveying was used in their construction, and celestially determined directions are a possibility. Curiously, in spite of numerous claims and frequent interest in the area, as yet almost nothing has been published and verified concerning such alignments. This unfortunate situation probably reflects the frequent doubts about whether certain structures were rebuilt in early excavation, about the lack of workers in the field, and perhaps about the almost overwhelming extent and diversity of the Anasazi ruins.

A. The Mesa Verde Sun Temple

Long suspected of astronomical orientation is the so-called Sun Temple, an enigmatic structure of mortared stone found atop a cliff in Mesa Verde, in the southwest corner of Colorado. It was first investigated by J. W. Fewkes, of the Smithsonian Bureau of American Ethnology, who found it through excavation in 1915.

Fewkes called the ruin a "temple" chiefly because its architecture and layout were unlike the Mesa Verde structures which were known to have been used for habitation and because, he felt, it was the work of a community effort. The adjective "sun" came from Fewkes's interpretation of a small pattern which looked like a rayed sun or the symbol for the sun and which he found etched in one of the stones at the exterior, southwest corner of the structure. (The "sun symbol" is now known to be a fossil imprint of a leaf.)

The Sun Temple is unusual in location and shape. Most of the associated Mesa Verde structures are in protected and inaccessible locations in cliff walls. The Sun Temple sits in an exposed location atop and near the edge of a mesa. This gives it an imposing appearance and the possibility of clear astronomical horizons.

The Sun Temple is a many-chambered structure which has the overall shape of a capital letter D. Fewkes measured the back of the D to be

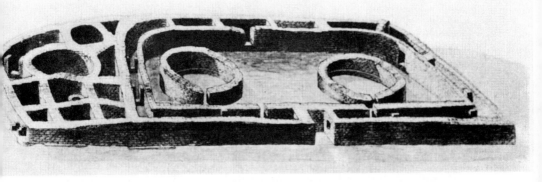

In 1915 J. W. Fewkes claimed that the straight wall of the Sun Temple at Mesa Verde National Park, Colorado, at the bottom of this 1916 sketch, was aligned with the summer solstice sunrise. A small fossil pattern resembling a sun symbol was found in stone at the lower left corner. (J. W. Fewkes, from Art and Archaeology, vol. 3 [1916], p. 341)

about 120 feet long and about 12 feet high at the highest point. From the pile of stone rubble in and around it, he surmised that the structure may have originally stood 18 feet high. Fewkes capped the walls with mortar, for preservation, and some believe that he may have rebuilt part of them—a point which is not answered by his available excavation notes.

Fewkes suspected that the straight side of the D was astronomically oriented to the rising point of the sun at summer solstice (and therefore to the setting point at winter solstice). Apparently he felt he had confirmed this through observation of the rising of the sun at the *equinox* (which defined the eastern point of the horizon) and by estimating the angle between the place of sunrise (east) and the real wall angle. He apparently did not determine the wall orientation with compass or transit, and, so far as we know, he did not make sunrise observations at either of the solstices.

One of those who in subsequent years had something to say about the astronomical alignment of the Sun Temple was L. J. Robinson of El Camino College in Torrance, California. Robinson claimed in 1955 that the straight wall of the D was aligned to the equinoctial rising points of the sun, that is, exactly east–west. This is practically the opposite of Fewkes's study, but it is not clear whether Robinson was aware of Fewkes's contradictory finding. Robinson also pointed out that there were coaligned holes in the interior walls of the structure, particularly along the straight side of the D, which were likely sighting ports and which would therefore have improved the precision of the building's original alignment. If the structure were really aligned to the equinox points, it would imply, of course, a rather sophisticated surveying method, for the east–west points are not nearly so easy to identify as the extremes of summer and winter solstice.

A subsequent and more thorough investigation was made by J. E. Reyman, an anthropologist at Southern Illinois University, and de- scribed in his dissertation in 1971. With a transit, Reyman determined that the tangent to the straight wall of the D was within 0.5 degree of the summer solstice sunrise/winter solstice sunset direction, apparently verifying Fewkes's original claim. Later, however, Reyman observed sunset at winter solstice at the site and found that the wall was not so aligned. He attributed the difference to possible error in the earlier transit measurements cited in his dissertation. This would seem to cast doubt on other possible astronomical alignments of the Mesa Verde Sun Temple which Reyman had mentioned in his dissertation. Among these were possible auxiliary markers for sunrise, midday, and sunset at the equinoxes and possible alignments on unspecified parts of the path of the moon. The clearest summary at this point seems to be that the alignment of the Mesa Verde Sun Temple is as yet unknown, or at least unpublished, after sixty years and a number of conflicting claims.

B.   Other Mesa Verde Sites

Reyman also described a number of other possible astronomical associa- tions at Mesa Verde. Included were wall alignments, window align- ments, a large body of data on kiva alignments, and descriptions of wall markings which may have been astronomically inspired. Several of the proposed alignments are for rising and setting azimuths of Venus. Since the declination of Venus swings through nearly 65 degrees, the appar- ent alignment of any structure on its place of horizon crossing at one arbitrary date is probably no more than chance. Venus, the most bril- liant of all the planets and stars, surely captured the attention of all primitive sky watchers. But any who noted its place of rising or setting for more than a night or two would have detected its shifting pattern which oscillates about the sun. Likely to be marked or noted would have been its places of heliacal rising or setting, when the planet would appear only momentarily at dawn or dusk, but these do not recur at simple fixed azimuths for the observer at any one place. Dedicated primitive astronomers might have marked the extreme azimuths of its rising and setting, which follow long-term cyclic patterns like the ex- trema of the moon. No evidence for this sort of sophistication has been found north of Mexico.

Reyman did find an interesting star pattern pecked in the ceiling above a north-facing window in the square tower of the Cliff Palace ruin. The pattern—seven dots which resemble a dipper—may represent the Little Dipper, or Ursa Minor, which was visible through the window. This association is based upon a simple window view, without definite fore- and backsight markers and hence is far from precise, as Reyman warned. The association, if real, would not be surprising. We can imagine that a familiar star pattern, seen out a window, might inspire one to copy it on a surface within reach. And although the stars of the Little Dipper are faint (about five times fainter, on the average, than those of the Big Dipper) its association with Polaris, the polestar, has made the Little Dipper one of the best known of all star groups. We should probably be surprised if it weren't recorded in some way by almost all primitive peoples.

C. Alignments at Chaco Canyon and Other Southwest Sites

Also investigated by Reyman and described in his dissertation are possible astronomical associations at the Aztec Ruins, Chaco Canyon, the Village of the Great Kivas, Gran Quivera, and several other Anasazi ruins in New Mexico and Colorado. For these sites, including Mesa Verde, he proposes possible star alignments for at least eighty-eight different kivas, great kivas, walls, towers, and other structures. The reality of the great majority of the associations is subject to considerable doubt. Most alignments are on dim stars which cannot be seen within several degrees of the horizon. Of the eighty-eight star associations, all but ten are on objects of magnitude two or dimmer and only three are from the ten brightest and most likely to be marked stars. Many of the proposed alignments are on circumpolar stars. These are very unlikely, for they never cross the horizon. Reyman has concluded that kivas show preferred directional orientations which are probably not astronomical, but architectural, for example, to provide good ventilation.

At Chaco Canyon in New Mexico Reyman found possible solstitial alignments in two major structures, Casa Rinconada and Pueblo Bonito. Casa Rinconada is a circular, walled structure with T-shaped doors and scattered wall niches. Like most other structures in Chaco Canyon, Casa Rinconada is built of dressed-stone bricks in thick walls which are laid with impressive precision and regular pattern. It is one of the so-called great kivas, which like the smaller ones were probably built for religious or ritualistic use. It lies partially sunken in the ground, in a broad, shallow valley between low, surrounding mesas which provide slightly elevated local horizons.

Reyman has proposed that the solstice sunrise and sunset points might have been marked at Casa Rinconada by foresights on the surrounding mesa top, for which he proposes signal fires. He assumed that the backsight was at some point in the kiva, perhaps the center of the structure. In searching the surrounding mesa top, Reyman found fire-burned areas in the directions of winter solstice sunrise and winter solstice sunset. Pottery found at these fire-burned areas was of a type contemporaneous with Casa Rinconada (c. A.D. 1100). No certain areas were found at the corresponding points of summer solstice. Other burned areas were found on the mesa top, but these were not considered. They did not fall on astronomical alignments.

Pueblo Bonito is probably the best known of the structures at Chaco Canyon. It is the well-preserved ruin of a large, D-shaped, multistoried communal dwelling which was occupied at about the same time as Casa Rinconada. Reyman has found two corner windows in remaining third-story rooms of Pueblo Bonito which appear to be aligned with the direction of sunrise at winter solstice. The precision of the alignments is not specified, and the base lines over which the alignments are proposed are only the thickness of the window jambs, that is, the thickness of the wall of the structure. Since the accuracy of any alignment depends upon the distance between foresight and backsight, the usefulness of a short base-line alignment, such as the one proposed, would be extremely limited. Furthermore, the possibility of chance alignment is high, for there were in Pueblo Bonito rooms and windows to accommodate five hundred to one thousand persons. Moreover, as Reyman points out, it is not certain that the two windows had unobstructed views at the time that the entire structure stood.

At Chetro Ketl, another major structure in Chaco Canyon, Reyman found wall alignments which were within 2 to 3 degrees of the direction of summer solstice sunrise (and winter solstice sunset). He considered these to be doubtful because of the possible effects of (modern day) reconstructions of the walls. Indeed, the persistent doubts of change in reconstruction haunt many of the proposed Anasazi alignments. These same problems plague analysis of the Sun Temple at Mesa Verde, and the archaeoastronomy of the American Southwest is still ambiguous.

Other scientists have looked into the possible alignments of the major structures at Chaco Canyon. In a recent report, R. A. Williamson, H. J. Fisher, and A. F. Williamson of St. John's College and Ranger Clarion Cochran of the Aztec Ruins and Chaco Canyon National Monuments reported surveying data which indicate astronomical alignments of three great kivas, although their findings differed from those made by Reyman. On the surrounding mesa tops they found directional markings cut into the rocks which were aligned with the direction of sunrise at winter solstice.

The remarkable Chaco Canyon ruins offer an almost irresistible temptation to amateur scientific speculation, and a number of claims have been made offering elaborate astronomical explanations for many features and structures. Some have noted intricate patterns of cross-kiva alignments by which rays of celestial objects illuminate specific wall niches of the great kivas at unique astronomical times; these purported alignments are then taken to indicate that the Anasazi Pueblo people at Chaco had advanced computational abilities and methods of predicting eclipses. There may be truth in some of these conjectures, but none has been presented quantitatively for astronomical verification, and none has been considered plausible by professional archaeologists who are familiar with other known aspects of the Chaco ruins, including the crucial matter of the known sequence of construction, use, and reconstruction of the buildings.

QUESTIONS OF THE EARTHEN MOUNDS

Best known of the structures left behind by the early inhabitants of eastern North America are the earth mounds, which are found chiefly in the valleys of the Mississippi and its tributaries. Many have been destroyed, but the known examples still number in the thousands. They differ significantly in size, form, era of construction, and, very probably, in use. Some are no bigger than a grave; others are a hundred feet high and many acres in extent. Some are geometrically regular and precise; others are simple heaps. Many were built as burial structures; others were probably not.

Serious investigations of the American mounds have destroyed the once-popular myth of a mysterious race of so-called Mound Builders, and modern archaeology attributes them to various groups of early-day agricultural and pottery-making Indians. The mounds were built, we now understand, by three successive groups or cultures starting perhaps three thousand years ago. The first, known as the Adena culture, left burial mounds in present-day Ohio, Indiana, Kentucky, West Virginia, and Pennsylvania. They also built more elaborate and more poorly understood mounds, including many in the shapes of reptiles, birds, and other animals. The best known of these is the Great Serpent Mound in Adams County, Ohio.

A later culture, the Hopewell, overlapped and then supplanted the Adena culture several hundred years before Christ. The Hopewell people continued to build mounds, chiefly in Ohio, Illinois, and through the Mississippi Valley. Some of the Hopewell mounds were elaborate burial mounds; others were impressive geometrical structures—perfect circles, squares, octagons, and straight lines. A good example is found in the well-known earthworks at Newark, Ohio, where an octagonal pattern, several acres in extent, is connected by a causeway to a circle of similar size.

Last, a culture known as Mississippian constructed massive platform mounds, frequently pyramidal in form, through the Mississippi Valley and adjoining regions of the present-day southern states. Some of the mounds were surmounted with buildings, dwellings, and temples of wood and thatch. The most elaborate and extensive of the temple mound sites was the small-city complex that was later called Cahokia, near Collinsville, Illinois, a few miles from East St. Louis. The Mississippian culture flourished from about A.D. 1000. Some of the North American Indians were still using platform mounds when European explorers first explored the continent in the sixteenth century.

There are good reasons to suspect that some of the mounds, in some or all mound-building eras, were associated with the sky, in orientation or purpose. Probably, too, the pyramid builders of Mesoamerica and the North American builders of the temple mounds were in contact, for some of the Mississippian mounds bear striking resemblances to Mesoamerican pyramids. There are evidences other than architectural similarities for this contact, and it is commonly proposed, if not accepted, that mound building in North America received at least some of its impetus from peoples to the south. As evidence grows for astronomical uses of certain of the Mesoamerican pyramids and platforms, so, by association, does suspicion grow that similar uses may have been built into some of the North American mounds.

Other bases for suspecting astronomical associations in the mounds are quite independent of Mesoamerican contacts. The regular and symmetric forms of some of the Hopewell mounds suggest sky alignments. The mound builders were an organized, agricultural people with need of a calendar reference. Finally, some of the effigy mounds represent figures which have counterparts in the Indians' panoply of constellations.

In discussing possible astronomical associations of the American mounds, we are in an area of almost complete conjecture. Little organized fieldwork has been done. Very few claims have been made for any specific associations. Even less has been published in the scientific literature. This is, however, an area where thorough and objective inquiry is due. Careful, documented measurements and skeptical interpretations, with the help of those who are professionally expert in the mounds and their known history, are badly needed. Proper tests for astronomical alignments and associations could add new information to the matter of possible Mesoamerican contact. If found, astronomical orientations would give insight into the level of culture of the builders of the mounds and, in certain cases, could provide an independent check on construction dates.

A. Mounds as Symbols of the Constellations

In recent work psychologist Thaddeus Cowan of Kansas State University has proposed that some of the conical mounds were built to represent individual stars, sometimes grouped in star patterns, and that certain of the effigy mounds represent constellations. The religion of the mound-building Indians centered on sky deities, and there are apparent connections between Indian sky legends and features of some of the mounds. The mounds were large and were sometimes built on hilltops, as though directed at the sky.

Cowan proposes that the Great Serpent Mound in Ohio is a representation of the Little Dipper (Ursa Minor, or Little Bear) and the polestar and that other bird and bear effigy mounds represent the Big Dipper (Ursa Major, or Great Bear) and the Northern Cross (Cygnus, or the Swan).

The physical similarities are subtle, and the associations proposed seem far from certain. The Great Serpent Mound is a long, twisting hill, about 20 feet wide, 4 or 5 feet high, and ¼ mile long. It looks very much like a snake with a coiled tail and with a round object, perhaps taken as an egg, in its mouth. Cowan regards the over-all curve of the twisting serpent as the handle and top two bowl stars of the Little Dipper; there is nothing in the mound which matches the Little Dipper's complete bowl, which, significantly, is brighter and more conspicuous than the handle. For the serpent to match the curvature of the handle of the Little Dipper, Cowan puts the polestar in the coiled tail. He points out that the direction of coil—a clockwise spiral—is the same as the apparent direction of diurnal rotation of the Little Dipper about the north pole. In addition, he cites a purported connection, in the symbolism of Indians of the Ohio region, between a serpent and the swastika—a common symbol which signifies rotation. Cowan does not claim perfect mapping between any of the effigy mounds and precise star patterns but bases his case largely on interpretations of symbolism, legend, and lore. Another large serpent mound, in Otonabee Township, Ontario, Canada, has an oval shape near its head, but its tail does not coil into a spiral. This second snake does not confirm the astronomical association.

B. The Newark Earthworks

The remains of the remarkable Newark Works, in central Ohio, are striking examples of Hopewell geometrical mounds and obvious candi-

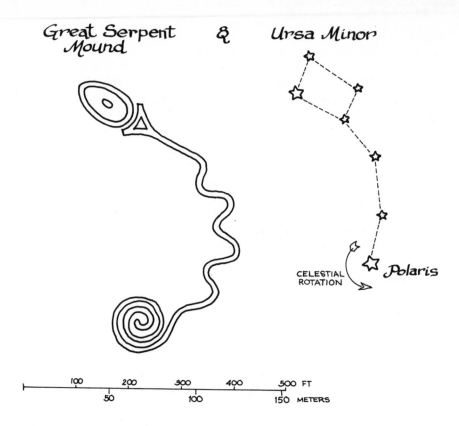

Thaddeus Cowan argues that the Great Serpent Mound in Ohio is a representation of the Little Dipper. (John A. Eddy)

dates for tests of possible astronomical orientation. The original complex, as found by early European explorers and settlers, covered about four square miles. Remaining today are the outlines of two circles and an octagon, traced out in broad and rounded earthen walls, from 5 to 15 feet high. One of the circles, about 1,200 feet in diameter, is connected to the octagon by a parallel causeway which is outlined in the same way. Together the circle and octagon cover about seventy acres, which fortunately are protected by the city of Newark as a park and golf course. I have found by simple map examination that a line through the axis of symmetry of the octagon, the center line of the causeway, and the center of the connected circle coincides very nearly with the direction of northernmost rise of the moon at the latitude of Newark. This extreme point of moonrise, which Alexander Thom calls the northern major lunar standstill and which he finds marked in many European Megalithic monuments, is reached in cycles of 18.61 years. Other investigators have found possible alignments at Newark on the

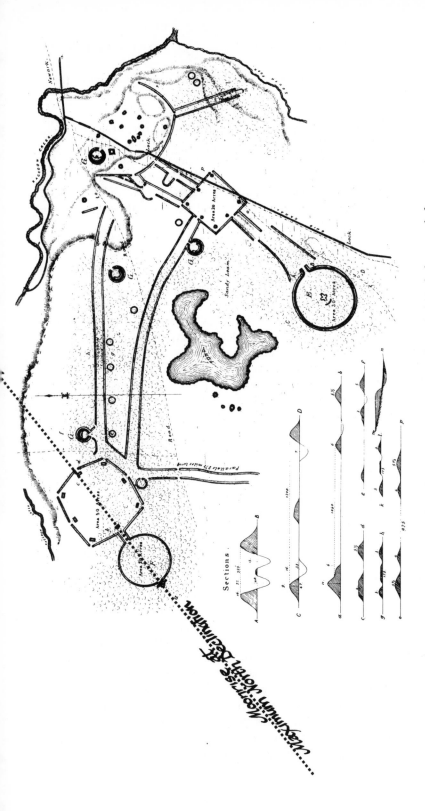

The connected circle and octagon at the upper left of this 1848 map of the earthen mounds at Newark, Ohio, cover an area of about seventy acres and are preserved today. The axis of symmetry of circle and octagon, by map measurement, coincides with the major standstill northern moonrise, an extreme reached every 18.61 years. (John A. Eddy)

rising of the summer solstice sun, but nothing has been published in the scientific literature.

C.   A Possible Woodhenge at the Cahokia Mound Site

The most elaborate of the known Mississippian mound-builder sites is Cahokia, near East St. Louis. At one time thousands of people lived there, in an elaborate and organized city, built around ceremonial centers of high, pyramidal earthen mounds. A central group, of over one hundred mounds, was occupied in the period between about A.D. 800 and 1550.

About eighty of the Cahokia mounds survive today. The largest, Monks Mound, is the largest Indian mound in America north of Mexico, and the largest prehistoric earthwork in the world. It is built on a base that covers sixteen acres and rises in four steps to a height of about 100 feet.

In archaeological work in the early 1960s, Warren Wittry of the Cranbrook Institute of Science, near Detroit, found evidence of four large posthole circles in an area that lies about one half mile west from Monks Mound. One of these circles he interpreted as a "woodhenge" (or circle of wood posts) which could have been used to mark the directions of sunrise at summer and winter solstice and at the equinoxes.

Only about half of this circle of postholes was found. From it, however, Wittry determined that it was part of a precisely drawn circle, about 410 feet in diameter, and that the full circle most likely consisted of forty-eight holes or pits (twenty of which were found). Each pit was about 2 feet wide, 4 feet deep, and elongated in one dimension to about 7 feet. Wittry concluded that the pits were this shape to ease the erection of large wooden posts. Each post was presumably laid on the ground with its foot in the elongated direction of the pit. In the raising, the post would naturally slide further into the hole, as into a socket. There is no way to establish the height of the posts that were used; indentations indicate that their diameters were about 2 feet.

From the samples that were found, Wittry established that the postholes were evenly spaced around the circle at increments of 7½ degrees. He found a posthole at each of the north and east cardinal directions, as measured from the center of the circle. Five feet east of the center of the circle was another posthole. Wittry found that this offset hole was in position to serve as an accurate solstice backsight. A line from the offset center hole to postholes on the circle in the northeast and southeast quadrants (foresights) marked the directions of sunrise at summer and winter solstice. The posthole at the eastern cardinal point would serve as a foresight for sunrise at the equinoxes. Since the foresight and backsight postholes were separated by about 200 feet the

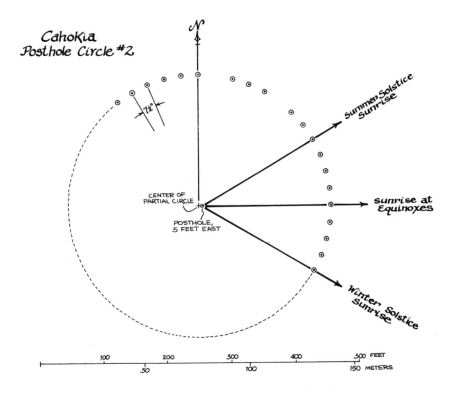

Cahokia
Posthole Circle #2

CENTER OF
PARTIAL CIRCLE

POSTHOLE,
5 FEET EAST

Summer Solstice Sunrise

sunrise at Equinoxes

Winter Solstice Sunrise

| 100 | 200 | 300 | 400 | 500 FEET |
| 50 | | 100 | | 150 METERS |

Warren Wittry found several posthole circles at the Cahokia mounds in southern Illinois, near East St. Louis, and in 1970 interpreted one of the circles astronomically. Identified postholes are shown as circled dots. (Warren L. Wittry)

alignment could have been precise enough to be of some practical calendar use.

In his paper on the subject, Wittry said nothing more about other possible alignments, such as the corresponding sunset directions or the lunar standstills. In such an elaborate structure we might expect to find these and possibly others. The postholes in the western part of the circle were not located and therefore nothing can be said with certainty about possible markings in that direction, although new excavations are imminent and should clarify the situation. If the holes were regularly spaced all around the circle (presuming the circle was complete), and if the same backsight hole were used, the offset hole east of the center hole would seem to throw off the sunset positions. The other circles Wittry found were also incomplete, and in several instances they intersected each other. All four posthole circles should be carefully resur-

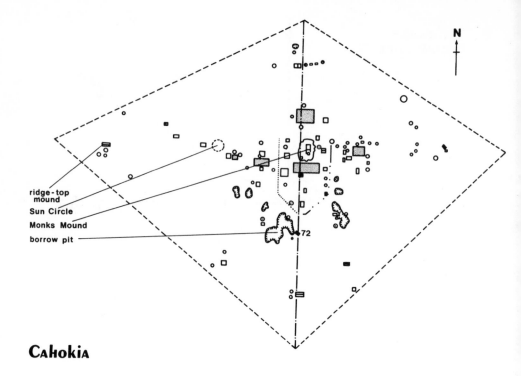

Cahokia

Melvin Fowler's map of Cahokia shows the alignment of the mounds and the entire urban complex with the north–south and east–west axes. Six of the mounds are ridge-top mounds and four of these are located at the corners, or city limits. According to Cahokia archaeologist Fowler, they may have defined the cardinal direction axes on which the city plan was based. (Griffith Observatory, after Melvin Fowler, from "A Pre-Columbian Urban Center on the Mississippi." Copyright © 1975 by Scientific American, Inc. All rights reserved.)

veyed to establish, if possible, the complete pattern. Wittry mentioned other postholes which were not parts of the circles, and these may have been detached foresights or backsights. Recently Wittry discovered the remains of an ancient cup in a pit in front of the winter solstice sunrise post. A design on the cup seemed to symbolize the cardinal directions and the horizon extremes of the winter solstice sun.

D. Council Circles on the Kansas Plain

Representing a later, different culture are a number of low, circular mounds in Rice and Cowley counties on the central Kansas plain which are known as the "council circles." In 1967 the archaeologist Waldo

A pair of mounds, one conical and the other flat, lies upon the main north–south axis of Cahokia. (E. C. Krupp)

Three council circles in central Kansas were left by early historic Plains Indians. Waldo Wedel, in 1967, reported that they are placed astronomically and that solstice alignments may be present. (John A. Eddy)

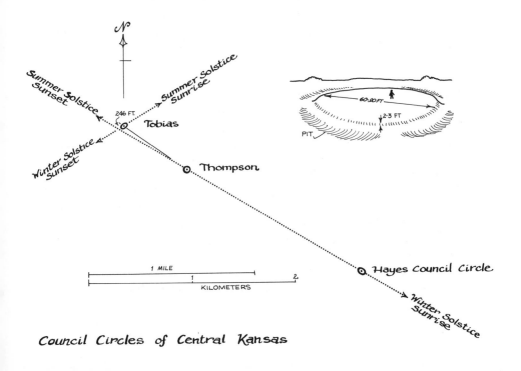

Council Circles of Central Kansas

Wedel of the Smithsonian Institution reported that two of these, a little more than a mile apart, appeared to have been built on a winter solstice sunrise line.

The council circles were presumably built by Plains Indians of the
Wichita tribe about three hundred years ago. They are found on sum-
mits of flat- or round-topped ridges. Each consists of a roughly circular
mound, no more than 3 feet high and 60 to 90 feet in diameter, sur-
rounded by a larger area of shallow, scooped-out depressions. The
depressions, which extend the circle diameters to 200 feet or so, are
presumably borrow pits from which dirt was taken to build up the low
central mounds.

Wedel reported five council circles, all in two Kansas counties, although he surmised that there had been others now destroyed by culti-
vation and land development. Each of the five is closely associated with
a known village complex, with one council circle per village. Wedel
proposed that the raised, circular mounds were the places of residence
of the elite of the Indian village and that the other Indians lived around
them, in part in the scooped-out basins. Excavation of the basin areas
confirms that they were used for habitation.

Three of the known council circles lie within a two-square-mile area
in Rice County, Kansas, between Wichita and Salina. Wedel excavated
one of these, the so-called Tobias Circle. He found that the other two,
the Thompson and Hayes circles, lay on a northwest–southeast line
which points to the sunrise at winter solstice. The same line in the opposite direction therefore points to the approximate place of sunset at
summer solstice, although Wedel did not comment on this. He determined the alignment direction by aerial survey, and associates
confirmed the winter solstice alignment of the centers of the Thompson and Hayes circles by on-site visual observation at sunrise on December 21 and 22, 1965. The distance between the two circles is about
6,400 feet, and they are plainly in sight of each other.

As further evidence that the circles were built on an intended alignment, Wedel noted that the Hayes Circle was elliptical in shape, with
its major axis along the line which connected it with the Thompson
Circle. He found that the third, the Tobias Circle, was also elliptically
extended, but in the direction of summer solstice sunrise. From these
findings, Wedel surmised that there could have been a common observation point from which a line to the Tobias site, 246 feet away, would
have marked the summer solstice sunrise, and, from the same point, a
line through the Thompson and Hayes circles (2,460 and 8,860 feet distant) would have pointed to the winter solstice sunrise. Wedel reported
no distinctive marking at the proposed intersection point, which leaves
the summer solstice (Tobias Circle) line highly conjectural. Moreover,

he did not give the measured azimuths of any of the alignments, and we do not know the uncertainties involved. A more detailed survey of the site positions is surely needed, to check Wedel's aerial survey and to investigate other possible interalignments of the three mounds.

ROCK PATTERNS ON THE WESTERN PLAINS

It is probably true that at the time of the arrival of Columbus in 1492 there were but a hundred thousand Indians on the western plains between the Mississippi River and the Rocky Mountains—probably fewer than the present population of Cedar Rapids or Colorado Springs. With so sparse a population distribution we should not be surprised to find how little is known and found of the early Indians of this wide region.

The most common relic of the early Indians of the western part of the Great Plains are tipi rings—circles of stones, typically 10 to 20 feet in diameter which were presumably used to hold down the edges of hide tipis. The rings are typically found in clusters, in places where tipis would likely have been pitched. Charcoal pits are often found in and near them, and carbon dating and lichen dating have shown that some tipi rings mark campsites more than a thousand years old.

A. *Medicine Wheels*

Less known and far fewer in number than the tipi rings are a scattering of stone alignments, effigy figures, and spoked wheels called "medicine wheels." Here the word "medicine" denotes magic or supernatural power and implies a possible association with social or religious ceremony. At least fifty western medicine wheels are known today, and they are found chiefly along the front (eastern) range of the Rocky Mountains, from Wyoming through Montana and into Alberta and Saskatchewan in Canada. Other examples, possibly related, have been reported as far south as southern Arizona.

Like the stone alignments and effigy figures, medicine wheels are simple patterns laid out on the surface of the ground with local rocks. They range in size from a few feet to several hundred feet across. Their defining characteristic is a set of spokes which radiate from a center or hub, so that, from above, each one looks like a primitive sun symbol or rayed sun. The number of spokes varies from wheel to wheel. Some wheels, but not all, have cairns, or piles of rock, at their centers or elsewhere in and around the patterns. Some of these cairns, as in the case of one large medicine wheel in Alberta, are as much as 30 feet in diameter and 6 feet high, containing perhaps 100 tons of rock and revealing considerable effort and motivation on the part of their builders. Other wheels are more casually laid and could have been built by a single person in a few hours.

It is probably significant that all of the known medicine wheels lie on hillocks, plateaus, mountaintops, and other elevated places with clear horizon views. This can be interpreted in several ways. The Plains Indians commonly sought out high places for isolated personal contemplation and vision quests, we are told, and these enigmatic patterns could represent their labors at these times. Or, if the wheels are simply expressions of art, as some interpret petroglyphs, it is surely the nature of artists to hang their work where it can best be seen, and prominent, high, clear areas of land would fit this urge for stone patterns. But the choice of clear horizon sites suggests another possible use for these largely unexplained structures: the horizon marking of important celestial events, such as the seasonal extremes of sunrise and sunset. The fact that medicine wheels resemble sun symbols adds weight to this possible interpretation.

I have investigated a number of medicine wheels for possible astronomical alignment: two good examples are the Big Horn Medicine Wheel near Sheridan, Wyoming, and the Fort Smith Medicine Wheel in southern Montana. Both show evidence of alignment on the sun at summer solstice. Since, in historic times, the Indians of this area held important ceremonies near the time of summer solstice—"when the sun is highest and the growing power of the world is strongest"—the alignments could have served for ceremonial calendar use. Another possible reason for marking or celebrating the summer solstice was agriculture. An assortment of Indian tribes moved through the regions where medicine wheels are found and some of these had relied upon agriculture either concomitantly with bison hunting or at an earlier phase in their history. Before the major population shifts and adjustments among the Indian tribes in the nineteenth century, domestic agriculture was important on the eastern plains, and it produced surpluses for trade throughout the region. While planting was probably never practiced in many of the rocky areas where medicine wheels are found, they may have been built there by people who had earlier agricultural traditions.

B.   The Big Horn Medicine Wheel

The Big Horn Medicine Wheel, which is probably the best known of all, reveals rather impressive astronomical alignments which suggest a thorough acquaintance with the sky. It is an elaborate wheel of twenty-eight spokes and roughly circular rim, found on a clear mountain shoulder at an altitude of about 9,600 feet in the Big Horn Mountains. Its diameter over-all is about 90 feet. At the center of the wheel is a rock

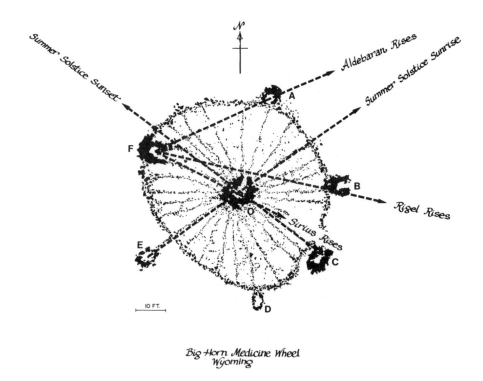

Big Horn Medicine Wheel
Wyoming

Astronomical alignments between cairns of the stone medicine wheel in the Big Horn Mountains near Sheridan, Wyoming, were reported by John Eddy in 1974. (John A. Eddy)

cairn about 12 feet in diameter and 2 feet high. Around the rim are six other cairns, one of which, in the southwest quadrant, is distinctive in that it lies well beyond the rim of the wheel on an extended spoke. All of the cairns are in the form of hollow, amorphous circles which make them consistent with places where posts might have been erected in the rocky soil. This is specifically true of the central cairn, beneath which a conical hole was found in the bedrock. The central cairn was probably used as a socket for a post or gnomon—the foot of the post set in the hole and tamped with loose dirt and the rocks of the cairn heaped around it for support. The other cairns, if used in a similar way, would have made the structure at one time a primitive woodhenge, although this is highly conjectural.

A line from the distinctive southwest cairn on the extended spoke through the center cairn points within about 0.2 degree to the place of rise of the sun at summer solstice. A line from another of the principal cairns to the center cairn marks the place of sunset on the same day, adding strength to the hypothesis of astronomical use, for if the dawn was cloudy, the sunset could have been tried. The possibility of chance alignment of two of six cairns on specific directions of sunrise and sunset at summer solstice is about one in four thousand.

Three of the remaining cairns of the Big Horn Medicine Wheel are positioned, by further chance or intention, to mark the points of rise of the three brightest stars of summer dawn: Aldebaran in Taurus, Rigel in Orion, and the brightest star of all, Sirius, in Canis Major. These stars, which we see so clearly in autumn and winter nights, rise just before the sun in summer. On certain specific days in summer they rise heliacally—that is, they appear at the horizon only momentarily and then fade into the light of dawn. The heliacal rising of the brightest stars provides a method of marking a calendar date which was used, we believe, by many early peoples. In the period between about A.D. 1400 and 1700 the one bright star in the sky whose heliacal rising marked the time of summer solstice at the Big Horn site was Aldebaran. Its marking by cairn alignments seems plausible and intentional, especially since archaeological dating of the site agrees with the use of the Big Horn Medicine Wheel in this period of time.

In the same period Rigel rose heliacally twenty-eight days after Aldebaran, and Sirius rose heliacally about twenty-eight days later still, in August, marking the end of tolerable weather on the mountaintop. It may be significant that there are twenty-eight spokes in the Big Horn Medicine Wheel, which might have been used for day counters.

And it may be completely unrelated, for the number 28 was a favorite of the Indian: There are 28 ribs in the buffalo. The number of days in a "moon," when precisely measured over long periods of time is 29½, which is close to 28; in primitive observation the difference would be subtle and hard to detect, since the exact time of new or full moon is uncertain by a day or so. Many reasons might dictate 28 spokes. It is my feeling, confirmed in part by archaeological investigation, that the spokes of the Big Horn Medicine Wheel are late additions to the structure and may have been added merely as a decoration and sun symbol.

C. Other Medicine Wheels

The astronomical alignment of the Big Horn Medicine Wheel has been known only since 1972, and as yet few of the other known wheels have been investigated in detail for possible similar alignment. The Fort Smith Medicine Wheel, on the Crow Indian Reservation in southern Montana, is a simpler structure with six spokes which radiate from a simple circular hub at the center. It lies on a flat hilltop and the spokes run down over the side of the hill. There are no associated cairns. If the

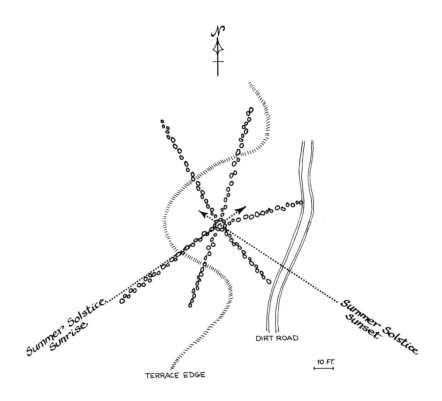

N

Summer Solstice Sunrise

Summer Sunset Solstice

DIRT ROAD

10 FT.

TERRACE EDGE

Fort Smith Medicine Wheel
Montana

The longest and most distinctive spoke of another, smaller medicine wheel, at Fort Smith, in southern Montana, aligns with the summer solstice sunrise. (John A. Eddy)

center of the Fort Smith wheel is taken as a foresight and marked with a post, the longest and most distinctive of the spokes points to sunrise at summer solstice, as confirmed by transit measurement and by observation at the time of summer solstice.

Thomas Kehoe, an archaeologist with the Milwaukee Public Museum, who has mapped a number of medicine wheels in Montana and Canada, has pointed out the striking similarity of the Moose Mountain Medicine Wheel in Saskatchewan with the Big Horn Medicine Wheel. Cairns of the Moose Mountain wheel appear in position to mark the same celestial features as those at the Big Horn wheel: sunrise at sum-

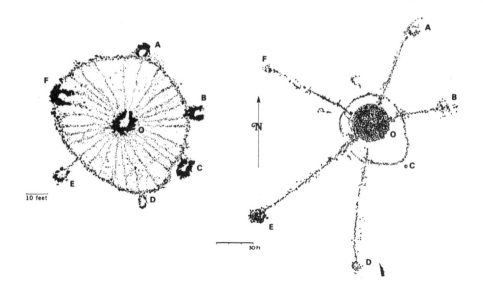

The Big Horn Medicine Wheel in Wyoming and the Moose Mountain Medicine Wheel in Saskatchewan, Canada, have similar arrangements of cairns. Both plans include alignments on the summer solstice sunrise and on three bright stars of the summer dawn. The Saskatchewan wheel is larger and also probably much older. (John A. Eddy)

mer solstice and the rising of Aldebaran, Rigel, and Sirius. A small group of rocks arranged in the form of a sun symbol, with short rays, was found at the end of the Moose Mountain wheel's summer solstice sunrise spoke. A recent survey on the site has confirmed the summer solstice alignment on sun and stars and suggests that the Moose Mountain site may predate the Big Horn wheel by one thousand years or more.

In earlier discussions with Indian informants of the Blackfoot tribe, Kehoe was consistently told that medicine wheels were built to mark the place of death of important chiefs. Some were known which had been built within the last century for this purpose. It seems likely that different medicine wheels, like mounds or other structures, were built for different reasons, by different people at different times. Or it is entirely possible that the same medicine wheel may have been used for different purposes. More of the known wheels must be investigated if we are to become convinced of their real origin and possible astronomical use.

D.   Rock Alignments

Human-made lines of rocks on the western plains are even less studied, harder to find, and less certain to interpret. Some are associated with bison pounds, to mark drive lanes. Many others are not. The author has investigated one of the recognized examples which lies in the Rocky Mountain National Park in northern Colorado, near Trail Ridge Road at an altitude of about 11,400 feet above sea level. Two lines of stones, each about 50 feet in length, meander roughly northeast and southwest downhill from a crude cairn on a rocky crest near an old Ute Indian trail. The stones are local magnetite and are sunken into the shallow mountain soil, indicating a certain antiquity. A sightline from the end of one of the two "spokes" to the center cairn (as a foresight) coincides with the local direction of sunrise at summer solstice. The line is not straight—although it may have shifted with the soil—and it is quite possible that the solstice alignment is purely accident. But it fits a pattern of summer sun alignments which have logical basis and which seem confirmed in a growing number of sites.

It is interesting to wonder why a solstice line might have been drawn at that high, cold, inhospitable place. No one who has driven the summit of the Trail Ridge Road—the highest highway in the United States —would ever propose that agriculture was practiced there. And any who step from their cars into the thin, biting winds of June will probably agree that only the masochistic would schedule celebrations there, at any time of the year. But I do not think we need invent practical reasons for every solstice marker, any more than we ask practical purposes for other religious or symbolic acts. I am prepared to think that an Indian who knew the simple secrets of the sun's recurrent path would find a certain pleasure in demonstrating his grasp of the sky by marking it in stones, sometimes in illogical places and for reasons that will be forever lost.

CHAPTER FIVE

Astronomy in Ancient Mesoamerica

ANTHONY F. AVENI

We know the peoples of ancient Mesoamerica did astronomy. They told us they did. Nearly all of the Maya writings, or codices, were burned by the Spaniards, yet those few that remain include extensive tables of the appearance and movement of Venus and the moon. Aztec picture books and accounts written shortly after the Spanish conquest refer specifically to astronomical observations. Glyphs on Maya stelae in the middle of the tropical jungle have been translated to reveal a complex and accurate calendar, based upon astronomical cycles and used throughout Mesoamerica.

Dr. Anthony F. Aveni, associate professor of astronomy at Colgate University, is one of the foremost experts on Mesoamerican archaeoastronomy. He, together with Professor Horst Hartung of the University of Guadalajara, Mexico, chaired the conference on archaeoastronomy in the New World, held in Mexico City in 1973 under the sponsorship of the Consejo Nacional de Ciencia y Tecnologia and the American Association for the Advancement of Science (AAAS-CONACYT). This unique conference and the compendium of papers Aveni edited from it, Archaeoastronomy in Pre-Columbian America, have stimulated considerable interest in and understanding of the role of astronomy in ancient Mesoamerica. Aveni has continued this effort with a second book on New World archaeoastronomy.

Even though enough of the written record remains to assure us that the Maya, the Aztecs, and others carried out systematic observations of the sky, there is no substitute for on-site surveys. Without an analysis of the ancient architecture, our knowledge of the astronomers' activities is incomplete.

Aveni has personally participated in and directed surveys at Teoti-huacán, Monte Albán, Uxmal, Chichén Itzá, and several other ruins he describes here. He has evaluated their astronomical potential and helped to clarify a subject which until recently was poorly understood.

Only within the last century have we begun to gain a full appreciation of the magnitude of the great civilization which developed on our continent before the arrival of Columbus. The eyes of the Western world were opened in the late 1830s, when John Lloyd Stephens, an American lawyer, was sent on a diplomatic mission to Guatemala and saddled with the task of seeking out the government, not a mundane assignment in Central America at the time. In his rambling, he visited numerous ruined cities, accompanied by a traveling companion, the talented British artist Frederick Catherwood. The brush of Catherwood and the pen of Stephens produced a magnificent two-volume work in 1841 entitled *Incidents of Travel in Central America, Chiapas and Yucatán.* Immediately a best seller, these volumes inspired the awe and imagination of every American reader at a time when the depths of our mysterious western frontier were being probed. Stephens destroyed the image of the indigenous red man as savage:

> Who were the builders of these American cities? They are not the works of people who have passed away and whose history is lost, but of the same races who inhabited the country at the time of the Spanish conquest.

Today's archaeologists tell us that great cities flourished between 500 B.C. and A.D. 1500 in Mesoamerica, in the region bounded by Arizona and New Mexico on the north and by Honduras and El Salvador on the south. Within lie several cultural nuclei: Olmec, Zapotec, Aztec, and Maya, the latter representing the pinnacle of artistic and scientific complexity in the New World.

In these pages we will be concerned with the astronomical achievements of these ancient people and, in particular, with the astronomical orientations in the architecture. In order to understand what may have motivated prehistoric Americans to build observatories and orient their cities astronomically, we begin with a brief examination of the native historical record pertaining to astronomical achievement, a record which, though scant, tells us what these ancient astronomers may have been watching and for what purpose.

THE WRITTEN RECORD

Father Diego de Landa, first archbishop of Yucatán, writing shortly after the Spanish conquest of Mexico, proudly tells us the sad tale of the fate of the written legacy of the ancient Maya of Yucatán. He refers to a great book burning which took place at the city of Mani:

> We found among them a great number of books written with their characters, and because they contained nothing but superstitions and falsehoods about the devil, we burned them all, which they felt most deeply, and over which they showed much sorrow.

Fragments of but four original Maya manuscripts or codices survive today. All contain a wealth of information relating to the heavens: lunar and solar almanacs, even a Venus calendar usable for one hundred years.

In Central Mexico the record is more complete. The primary role of the astronomer is illustrated in the Aztec *Codex Mendoza,* a document produced at the time of the conquest. The drawing depicts a high priest who, according to the caption provided by a native informant, is "watching the stars [eyes] at night in order to know the hour, this being his official duty." So important was this function that even Montezuma, king of the Aztecs, was required to arise "after dusk, at about 3 A.M., and immediately before dawn" to offer incense to certain principal stars. The commentary is by the historian Alvarado Tezozómoc.

The Spanish historian Juan de Torquemada, writing a century after the conquest, tells of the astronomical pursuits of Netzahualpilli, the king of Texcoco:

> They say he was a great astrologer and prided himself much on his knowledge of the motions of the celestial bodies; and being attached to this study, that he caused inquiries to be made throughout the entire extent of his dominions, for all such persons as were at all conversant with it, whom he brought to his court, and imparted to them whatever he knew, and ascending by night on the terraced roof of his palace, he thence considered the stars, and disputed with them all on difficult questions connected with them. (*Author's translation.*)

From these statements it is difficult to grasp the cosmological point of view espoused by these ancient sky watchers. We know that Mesoamerican thinkers conceived of a layered universe, each stratum containing one type of celestial body. Above the layer of the earth, the moon traveled its heavenly course. Above this lay the clouds, the stars, the sun, Venus, comets, et cetera, the male-female creator god occupying the thirteenth and uppermost layer. The underworld consisted of nine divisions, stacked in an orderly fashion below earth. This view is quite in contrast with both the geocentric (earth-centered) and helio-

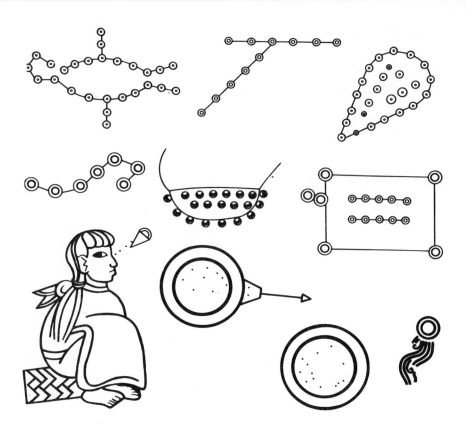

Astronomical drawings from different Aztec codices are shown in this draw-ing. A priest shown watching the stars at night is from the Codex Mendoza. *Several representations of constellations from the* Florentine Codex *are shown above. The figures in the lower right may represent two observatories and a comet.* (Griffith Observatory, after Anthony F. Aveni)

centric (sun-centered) views of the universe which evolved in ancient Greece. In the Mesoamerican universe, the place of man seems to be given no added importance over the rest of the system. Indeed, the con-cept of the center of the universe is not even suggested. The hierarchi-cal structure of the system is the basic theme.

For these people the primary interest in the course of the heavenly bodies lay in the utilitarian realm of timekeeping, mostly for astro-logical purposes. They may have made precise measurements, since Torquemada goes on to say:

I have seen a place on the outside of the roof of the palace, enclosed within four walls only a yard in height, and just of sufficient breadth for a man to lie down in; in each angle of which was a hole or perforation, in which was placed a lance, upon which was hung a sphere; and on my inquiring the use of this square space, a grandson of his, who was showing me the palace, replied that it was for King Netzahualpilli, when he went by night attended by his astrologers to contemplate the heavens and the stars; whence I inferred that what is recorded of him is true; and I think that the reason of the walls being elevated one yard above the terrace, and a sphere of cotton or silk being hung from the poles, was for the sake of measuring more exactly the celestial motions . . . (*Author's translation.*)

Pictorial elements excerpted from various codices strongly suggest that special pyramid-temples were used as observation posts. In one we see a priest perched in a temple doorway shown in profile. He seems to be peering over a crossed-stick sighting device as if to mark an event occurring at the local astronomical horizon. The outside of the temple is studded with star symbols. Is the stick a measuring device? What horizon event is being witnessed? Does the star temple have a special orientation toward the object on the horizon?

The eye-stick symbolism is repeated in two other codex drawings. The stick may have been used to sight a specific planetary or solar position along the horizon. When the object sighted returned to its original place in the notched stick, it completed its cycle.

Of all the celestial bodies observed by the Mesoamericans, Venus was among the most important. Called the Great or Ancient Star, it is repeatedly mentioned in the written record. The sixteenth-century priest-historian Fray Bernardino de Sahagún tells us that "when this star made its appearance in the east, they [the Aztecs] sacrificed captives in its honor, offering blood, flipping it with their fingers towards this star."

On pages 46 to 50 of the Maya *Dresden Codex,* we see a complete record of the apparition of Venus as morning and evening star. Several Venus years are recorded in the dot and bar mathematical symbolism of the Maya. These intervals represent the elapsed time between successive heliacal risings or initial appearances of the planet as morning star. The Maya divided this 584-day period into four subintervals, or stations, representing the appearance of Venus as morning star and evening star and its disappearance intervals in front of and behind the sun. These appear in the four columns A through D at the bottom (line 26) of the Venus table. Accompanying pictures show the Venus god of the Maya, Kukulkan, in several evil manifestations, spearing his victims. The spear may symbolize the rays of the bright planet as it first shines in the predawn sky. Evidently the heliacal rising of Venus bode evil for

Many examples of astronomical symbolism can be found in Mesoamerican codices. A circle with a chord drawn across it and with a smaller circle inside represents a star and may designate a building as an observatory. Many of these same representations of buildings and platforms include a pair of crossed sticks, which may have been an astronomical sighting device. (Griffith Observatory, after Horst Hartung and Zelia Nuttall)

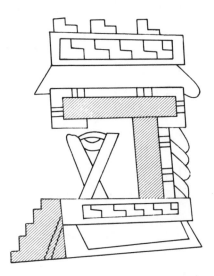

the people of Yucatán; therefore, they were quite concerned about where and when it might first appear after being hidden behind the sun. An added page of the codex was fashioned to serve as table of corrections to alter the Venus observations for use by astronomer-priests at a later time.

The Maya were conscious of the commensurability of the Venus year and the so-called vague year of 365 days, which they also employed in their calendar, and this may have induced them to watch Venus so closely. Eight vague years are exactly equal to five Venus years. The equivalent number of days is 2,920 and it turns up repeatedly in the Venus tables. After a cycle of 2,920 days, Venus returns to the same place in the sky very nearly at the same time of the year. The neat fitting together of the earth and Venus years may have induced the number-conscious Maya to watch the planet.

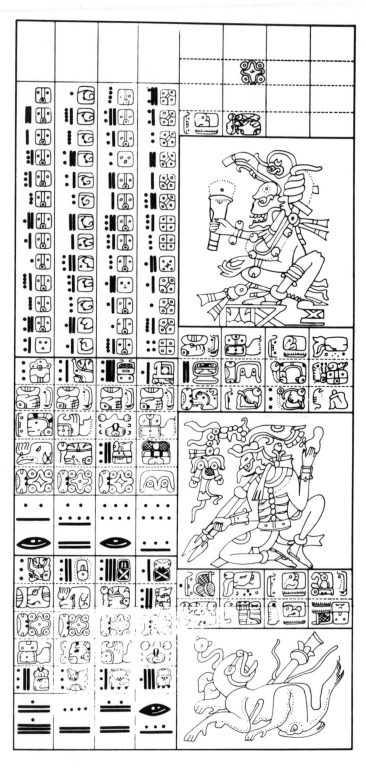

Maya Venus Table
Dresden Codex

The glyphs and numbers in the lower left-hand corner of the page of the Maya Dresden Codex *are shown schematically to comprise a portion of a table of positions of the planet* Venus. (Upper: W. E. Gates; lower: Griffith Observatory, after W. E. Gates)

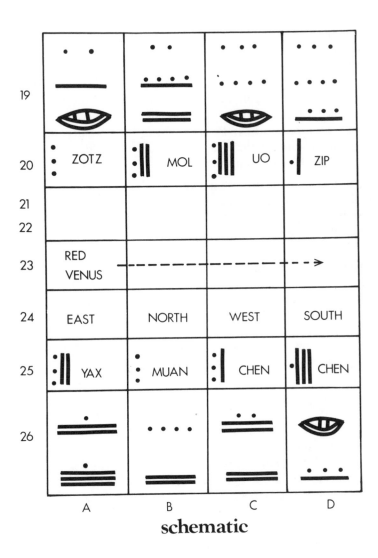

schematic

The Pleiades star group, known to European observers as the Seven Sisters, also received considerable attention in Mesoamerica. It found use in the setting of both religious and agricultural calendar dates. In the Aztec *Florentine Codex* of Central Mexico this group can be identified as Tianquiztli, or the Marketplace; among the Maya they were Tzab, the Rattle (of the rattlesnake). The Sherente of Brazil still use the Pleiades today to mark the seasons. They say that the year commences in June when the Pleiades first appear as the sun leaves the constellation of Taurus. This is taken to be a sign of wind which ushers in the rainy season. The Navajo dignify the Pleiades by associating them with their principal deity, Black God, creator of fire and light. In fact, a recognizable form of this group appears on the face of Black God.

The Aztecs determined the occurrence of their most important feast day by the appearance of the Pleiades. According to Sahagún, the ceremony of the Binding of the Years took place every fifty-two years and began when the Pleiades transited the zenith at midnight:

> And when they saw that they had now passed the zenith, they knew that the movements of the heavens had not ceased and that the end of the world was not then, but that they would have another fifty-two years, assured that the world would not come to an end. (*Author's translation*.)

Michael Coe, an archaeologist at Yale and a well-known scholar of ancient Mesoamerica, has examined in detail the attraction of celestial objects for the astronomers of ancient Mexico, and on the basis of his studies he has identified some interests which typified the approach of the native American astronomers. They were, it seems, deeply committed to the observation and manipulation of the orbital cycles of the moon and the planets as seen from the earth. Although a planet like Venus requires only 225 days to complete a circuit around the sun, from our moving observatory, the earth, Venus appears to take 584 days to go from its extreme position, say east of the sun, back to that extreme again. This second period of the planet is called the "synodic period" of the planet, and it is a very obvious and meaningful interval. The ancient Mesoamerican astronomers were alert to these synodic periods and incorporated them into their calendrical calculations.

The ancient Mesoamerican calendar was the product of many centuries of development and refinement. Its basic elements can be traced as far back as the Olmec people, who established ceremonial centers on Mexico's Gulf Coast at least as early as the second century B.C. The calendar became an intricate and involved system of timekeeping in which several different cycles of time were simultaneously used to measure the passage of time. Although the Mesoamerican view of time was not mechanical, their calendar systems are often described in terms of a set of interlocking gears in order to facilitate an understanding among us inhabitants of the twentieth century of a sense of time that was distinctly different from our own. In the 260-day sacred count calendar, for example, each day had a name and a number. Thirteen numbers and twenty names were available, and the number-name compounds followed the sequence obtained by imagining two interlocking gears. One of these has thirteen numbered teeth, and the other has twenty named spaces between teeth. Each day both gears are permitted to turn one division, and the compound name for that day comprises the number and name of intersecting tooth and space. The sequence would start with the day "1 Imix." The next day would be "2 Ik," and the third day would be "3 Akbal." In the first cycle of thirteen days only thirteen names would have been used, and so the fourteenth day would start with 1 again but with an as yet unused name, in this case, "1 Ix." The day name "Imix" would be reused in another seven days but this time in conjunction with the number 8 ("8 Imix"). The two cycles continued until all number-name combinations (260) appeared. The last day of the cycle was "13 Ahau."

Other cycles of time were also measured, and the Mesoamericans were interested in the various combinations of dates obtained from several different systems. Here again a complex, interlocking array of gears, one for each system, is a convenient analogy, for it illustrates what, in effect, the Mesoamerican calendar keepers were trying to keep straight.

According to Coe, the calendar keepers were interested in numerological relationships and carried out extensive calculations which related one time cycle to another. In this sense the ancient Mesoamericans were more like the Babylonians, who were similarly interested in numerical calculations, than the Greeks, who approached astronomy from a geometrical point of view.

Evidence, particularly from the codices, suggests that directions along the horizon also were important to the Mesoamerican astronomer-priests and that architectural modifications may have been made in certain buildings to emphasize the relationship between human beings and the cosmos.

AN AZTEC EQUINOX

A reference in the literature from the time of the conquest tells that the act of orienting a building astronomically was actually practiced. Father Toribio Motolinía, writing while among the Aztecs, tells us that:

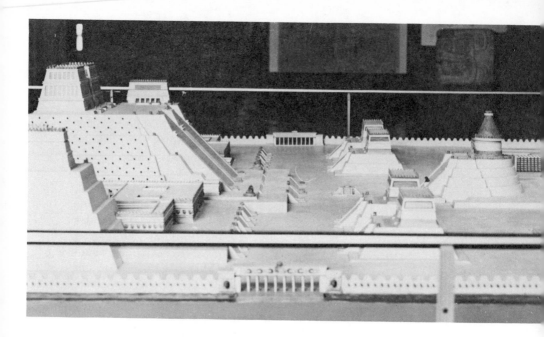

The massive pyramid of Templo Mayor in Tenochtitlán in eastern Mexico, seen in this model in the Museo Nacional Antropología in Mexico City, was topped by twin temples. Across the plaza and to the right is located the cylindrical Temple of Quetzalcoatl. (Horst Hartung)

The festival called Tlacaxipehualiztli took place when the sun was in the middle of Huicholobos which was at the equinox, and because it was a little out of the straight, Montezuma wished to pull it down and set it right.

In this passage Motolinía is speaking of the Templo Mayor, the principal religious structure in the ceremonial center of Tenochtitlán, the Aztec capital, and the pair of temples which surmount it. Evidently, the priest and worshipers faced east, probably on top of a small round structure, where they could view the sun rising at the equinoxes in the space between the temples. In an early map of the city sent to the king of Spain by Hérnan Cortés we actually see the sun pictured in the space between the oratories.

During a field trip to Mexico in 1974 the author and his associates located the ruins of the ceremonial center, the shattered remains of which lie submerged under modern Mexico City. Only the southwest corner of the temple is visible today as a broken wall a few yards high.

Using a surveyor's transit, we were able to measure the orientation of the western front of the building. We found that it was turned 7½ degrees clockwise from the cardinal directions. On March 21 and September 21 the sun rises at the east point of the horizon, but in the low northern latitude of Tenochtitlán it progresses slightly southward as it rises higher in the sky. Thus, it would appear in the notch between the temples in a direction 7½ degrees south of east. Using the archaeological plan of Tenochtitlán, together with our measurement of the skewed axis of the temple, we were able to calculate the height above the observation point at which the equinox sun would have been positioned exactly 7½ degrees south of east. The result, about 55 yards, is in good agreement with the height of the Templo Mayor based on purely archaeological evidence. Thus, both astronomical and textual evidence combine to identify the Templo Mayor as a functioning astronomical observatory.

Templo Mayor was aligned slightly south of due east of the Temple of Quetzalcoatl to permit observation of the equinox sunrise from the circular tower. (Griffith Observatory)

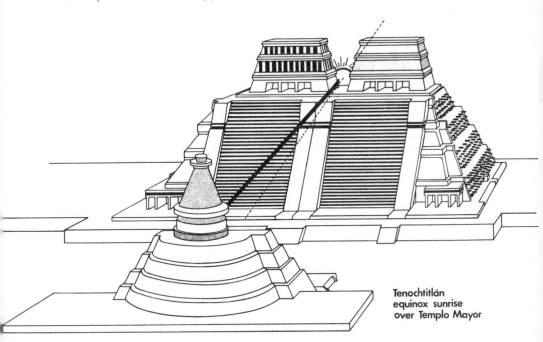

Tenochtitlán
equinox sunrise
over Templo Mayor

ASTRONOMICAL URBAN PLANNING

How widespread was this practice of orienting buildings? Our field trips have taken us to several dozen sites which we have now measured carefully. In most cases we find a right-angle grid structure underlying the plans of ceremonial centers. Peculiarly, the principal axes are nearly always skewed clockwise from the cardinal points on the horizon. In a distribution relative to true north of the principal axes of Mesoamerican sites which have been measured to date, 88 per cent of the sites studied are skewed to the east, rather than west of north. Few sites are found to be laid out on a perfect north–south axis. Instead, the diagram is centered about a direction 15 to 20 degrees east of north. One group of cities, at the peak of the curve, is called the "17-degree family of orientations." Another group, oriented at 7 degrees, may also exist. Let us concentrate on the 17-degree group which seems more obvious and can be localized in the region of Central Mexico.

The builders of Teotihuacán, largest and most influential of all cities of ancient Mesoamerica, may have been the originators of the 17-degree family. Built about the beginning of the Christian era on a grid which does not conform to the local topography (even the course of a river was diverted to fit the preordained pattern), the city possesses a major axis defined by the Street of the Dead, which is skewed 15 degrees 25 minutes to the east of north. Careful studies by the Teotihuacán Mapping Project reveal at least three different axes closely aligned about this direction. Researchers working with the project also discovered a possible baseline used in laying out the city. A design consisting of a pair of concentric rings and a cross was hammered into the limestone base of one of the buildings in the so-called Viking Group, adjacent to the Pyramid of the Sun and near the center of the city. (This group of buildings has nothing to do with Vikings but is named after the foundation that paid for the excavations.) About 2 feet in diameter and consisting of holes spaced an inch or so apart, it is still visible to tourists as it lies unprotected. Nearly two miles to the west on the slope of Cerro Colorado, project workers discovered a cross of similar size and shape carved into a basaltic outcrop. We measured a bearing of 15 degrees 21 minutes north of west for the Cerro Colorado cross as viewed from the cross in the Viking Group. This east–west baseline lies within a few minutes of arc of a right angle to the Street of the Dead. The Teotihuacános could have employed the baseline to lay out the principal axis of the site. Evidently, they constructed a very accurate right angle to lay out the Street of the Dead.

We tested the baseline for astronomical orientation by determining, for the latitude, building time, and horizon elevations of Teotihuacán, which astronomical bodies could have been viewed along either direc-

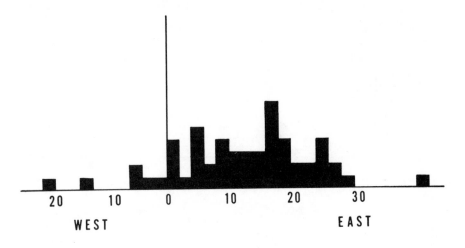

The alignment directions of measured Mesoamerican sites are plotted horizontally in this diagram, where 0 is zero degrees azimuth, or true north.
Peaks build up where a number of sites share the same orientation with respect to north, and these may indicate "families" of shared alignments. The
most prominent "family" in this distribution of fifty-six sites is the one centered on 17 degrees east of north, where a high peak can be seen. This represents the so-called "17-degree family," which includes Teotihuacán. (Anthony F. Aveni)

tion from cross to cross. Of the several possibilities the most likely
seems to be the Pleiades. Not only did this conspicuous star group set
within 1 degree of the east–west axis of Teotihuacán, but also the group
functioned in a most unusual way at that place and time. The Pleiades
underwent heliacal rising on the same day as the first of the two annual
passages of the sun across the zenith, a day of great importance in
demarcating the seasons. The appearance of the Pleiades may have
served to announce the beginning of this important day, when the sun
at high noon cast no shadows. Writing in the first decades of this century the well known scholar of Ancient Mesoamerica Zelia Nuttall associated the zenith passage event with the so-called Diving God, who
appears as an inverted figure on the friezes of so many Mesoamerican
buildings (including those at Teotihuacán). She writes:

> Twice a year shadows disappeared at noon as the Sun God descended.
> His descent was always followed by rains caused by the heat of the verti-
> cal solar rays. This momentary descent, which marked the advent of the
> rainy season, was of trancendental importance to the native agriculturist.
> After the descent of the god, they could confidently sow the seeds of
> maize and other plants with a certainty of rain.

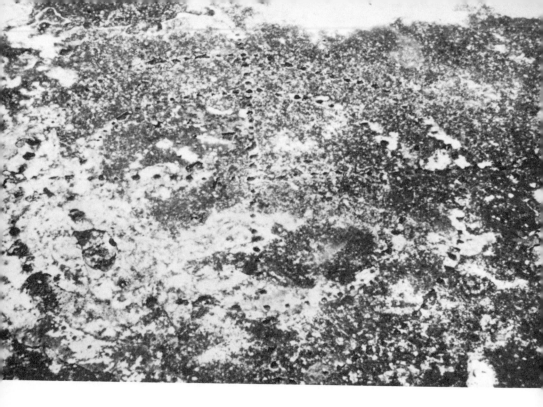

A cross is pecked onto the floor of a small room in the Viking Group at the center of Teotihuacán (top). A similar cross is pecked into the rock on the hillside of Cerro Colorado, to the west (below). (Anthony F. Aveni)

The main axis of Teotihuacán, in central Mexico, is the Street of the Dead, here viewed from the Pyramid of the Moon, looking south. (E. C. Krupp)

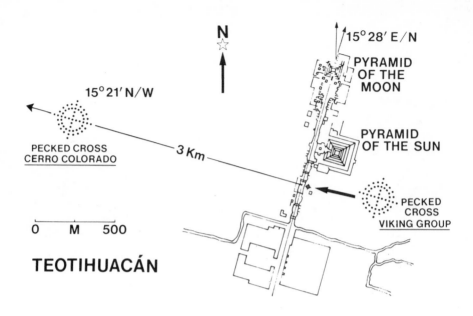

A possible astronomical baseline at Teotihuacán, perpendicular to the Street of the Dead, is defined by the two pecked crosses. (Griffith Observatory)

A photograph, made with the aid of the University of Florida Planetarium, shows by simulation the Pleiades setting on the western horizon of Teotihuacán in A.D. 150 as viewed from the pecked cross near the center of the city. The arrow points to a large tree behind which the Cerro Colorado cross is situated. A clear view of the horizon is obtained. While priestly observers stationed themselves in the center of the city, perhaps another group lit a fire at the base of the other cross to provide an accurate marker for the setting point of the Pleiades.

In view of both the coincidence of events and the central importance of the Pleiades in Mesoamerican star lore, this star group must remain the prime candidate for an astronomical motivation in the orientation of Teotihuacán. Nearly identical pecked-cross petroglyphs have been found at Alta Vista, over three hundred miles north of Teotihuacán and very close to the Tropic of Cancer, and at Uaxactún, a Maya site over six hundred miles to the south in the Guatemalan Petén jungle. The role of these crosses in the architectural planning at these sites is yet to be assessed.

Ceremonial centers in the vicinity of Teotihuacán, built as much as fifteen centuries after it, reflect the same sacred orientation. Included in this group are the pyramid of Tenayuca nineteen miles to the south-

The Pleiades are seen, in the sky at the left, to set over the western horizon at Teotihuacán in this simulation photographed in the University of Florida Planetarium (courtesy J. Carr, director). The Cerro Colorado pecked cross is behind the large tree (arrow). (Anthony F. Aveni)

west, the Casa de Tepozteco sixty-two miles to the south, both built immediately before the conquest, and Tula, the Toltec capital forty-four miles to the northwest. Table 1 offers a comparison of all the orientations we found at several Central Mexican sites. Those in the 17-degree family head the list. All lie within sixty-two miles of Teotihuacán, and in each case the orientation is east of north.

Although there seems to be no systematic relationship between the orientations of the Central Mexican sites and the dates of their construction, the table clearly establishes a tendency for the inhabitants of Central Mexico to skew the axes of their buildings to the east of north.

Tenayuca, Tepozteco, and Tula seem to have copied Teotihuacán in its orientation. Because of the movement of the earth's pole of rotation among the stars, by the time these centers were erected the Pleiades no longer set along the Teotihuacán east–west axis, nor did they announce the zenith passage of the sun by the tenth century A.D., a time when the great civilization of Mexico had passed its prime and was well on the decline. The priest rulers of these new centers probably looked upon Teotihuacán as a holy city in ruins. Out of reverence for the past, they may have planned their centers of worship with the same directional axes. The sacred alignment could have been transferred astronomically simply by sighting a substitute star which had replaced the Pleiades. By postclassic times that star group set several degrees to the north of "Teotihuacán west." Thus, we view the axes of Tula and the other members of the 17-degree group as non-functional imitations of

TABLE 1

Orientation of Major Axes of Central Mexican Sites

Site	Orientation (East of North)	Approximate Building Date (Century: B.C. =—; A.D. = +)
Teotihuacán	15°25'	+1
Tenayuca	17°05'	+15
Tepozteco	18°00'	+15
Tula	15°49'	+10
Xochicalco	0°35'	+10
Teopanzolco	0°43'	+15
Calixtlahuaca	1°31'	+14
Cuicuilco	5°43'	−5
Tenochtitlán	7°24'	+15
Tlatelolco	9°32'	+15
Tenango	13°33'	+13
Chalcatzingo	19°19'	−9?
Cholula	26°16'	+12
Manzanilla	27°04'	+12

Teotihuacán. By postclassic times, the original purpose of the orientation was probably lost completely.

Buildings at Chichén Itzá, over nine hundred miles away to the east of Teotihuacán, show a similar plan. Comparison of the ground plan and orientation of the Toltec capital, Tula, with those buildings at Chichén Itzá that were built shortly after the Toltecs conquered the Maya of Chichén Itzá, around the tenth century, shows a relationship between the two sites. Not only are the axes of the two cities identical, but also the ballcourts, rectangular, walled structures in which the ancient ball game was played, are in the same relative position. What was the nature of the Toltec mind which caused them to transplant their orientation to a conquered city hundreds of miles away from their capital?

In 1974 geographer Franz Tichy published the results of an aerial survey he made of the highlands of Mexico. He found that villages and towns built after the Toltec conquest aligned to the east of north. The 17-degree family was detected in the survey. Apparently, later structures continue to preserve directions already important in antiquity. We are left to wonder whether the later architects were conscious of what they were copying.

We have suggested two possible astronomical phenomena which may underlie the east-of-north skew: the equinox sunrise at Tenochtitlán and the setting point of the Pleiades at Teotihuacán. It is unlikely that every orientation in Table 1 can be related to these events. Nevertheless, by varying the height of the observer relative to the horizon event of sunrise at the equinoxes, we can account for most of the orientations in the table, especially those in the 0–10 degrees east-of-north range. Alignments with a larger skew could be explained by a sunrise or sunset position on one of the significant agricultural, civic, or religious dates of the year, for example, the date of passage of the sun across the observer's zenith. Still, some problems relating to the east-of-north skew remain unsolved.

ARROWS TO THE SKY

Individual buildings with peculiar shapes or orientations at Mexican sites may have astronomical significance, too. Such buildings could have been distorted or thrown out of symmetry deliberately to emphasize a horizon event of importance. One of the most unusual buildings in this regard is Building J at Monte Albán, an elevated site used by the ancient Zapotecs near Oaxaca City. Appearing somewhat like home plate on a baseball diamond, its arrow-shaped ground plan is plainly visible in an aerial photograph. The Mexican archaeologist Alfonso Caso, who excavated the site in the 1930s, remarked the following:

> While all the other buildings have their axes directed sensibly toward the cardinal points, Building J, in contrast, has its stairway facing to the northeast. Also . . . its plan does not present a quadrangular or rectangular form as in the other mounds, since the back part ends in an angle.

Actually, the "back part" is nearly a right angle.

Caso suggested that a horizontal tunnel cut through the rear (south-western) portion of the building was probably used for making astronomical observations. Thus evolved the popular myth that Building J was an observatory. On the first of my visits to the site in 1970 I was surprised to learn that no one had ever measured and tested the "observatory" to see how it might have functioned.

Bas-relief carvings on the western wall of the building show the cross-stick symbolism clearly. Is this the same sighting device depicted in the codices and does its appearance suggest that the front of the building was skewed deliberately to face a particular astronomical event on the northeastern horizon? As we measured the direction of the perpendicular to the doorway of Building J, which is all that remains at the top of the building, we imagined the astronomer-priest seated there twenty centuries ago waiting for a particularly interesting celestial ob-

Building J at Monte Albán, in southern Mexico, is a peculiar, arrow-shaped structure. Building P is across the plaza, to the right, and the inked arrow indicates the position of the entrance to Building P's zenith sighting chamber. (Robin Rector Krupp)

ject to rise above his horizon. Did the event mark the beginning of a special cycle in the calendar or a sacred festival? Was it evil or good?

We laid out a baseline perpendicular to the doorway to find the exact direction of the axis of the supposed observation chamber. The view looked out over the northeast horizon in a direction 47½ degrees east of astronomical north. Our computer told us that the only significant astronomical event occurring within a ½ degree of that direction in 275 B.C., the time of construction of the building according to radiocarbon dating, was the rising of Capella, sixth brightest star in the sky. We were especially surprised when the computer told us that Capella was unique among the bright stars at that time, since it underwent heliacal rising precisely on the same day as the first annual passage of the sun across the zenith of Monte Albán, which was May 9 at the time. More clues to this cosmic-architectural puzzle were revealed by a careful study of the plan of the ruins. Our doorway perpendicular cut across Building P on the east side of the plaza. Halfway up and in the center of the stairway of that building, precisely on line with the perpendicular, we discovered the entrance to a chamber containing a narrow vertical shaft drilled upward to the overlying steps. Peering through the opening we could see a spot of blue sky at the zenith. The shaft evidently served as a sight tube to observe the zenith passage of the sun or perhaps the Pleiades which also transited the zenith. The connection between Buildings J and P was further accentuated by a perpendicular to the stairway of J which passed through the doorway of P.

What is the most likely solution to this puzzle? Capella, by first appearing in the early morning sky, announced the day of zenith passage in Monte Albán. When this happened, the priests could descend into the passage in Building P to make their solar observations. These cosmic events were so important that the Zapotecans endowed the earthly realm with their permanent presence by skewing Building J to line up with the zenith tube in Building P, thus making a sort of architectural calendar.

For the sake of completeness, we examined other astronomical alignments which might pertain to the configuration of Building J. The tunnel discovered by Caso served no obvious function. Sun and moon positions bore no relation to any portion of the structure, but the "point" of the arrow at the southwest side of the building was directed close to the setting position of five of the twenty-five brightest stars in the sky (three stars of the Southern Cross and Alpha and Beta Centauri). Though these stars set far to the west of the arrow point today, in 275 B.C. each set within 2 degrees of the point and within a quarter degree of each other. These events served no particular function, unlike

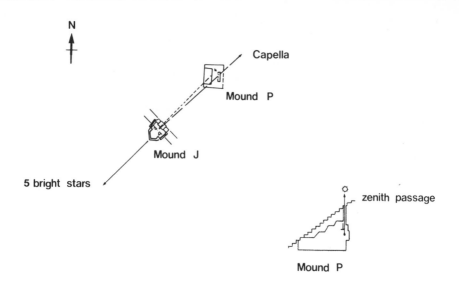

Monte Albán Mound J

The perpendicular to the ruined doorway on top of Building J at Monte Albán points northeast and aligns with the rising point of Capella. Capella's heliacal rising coincided with the year's first zenith passage of the sun in ancient Monte Albán. The perpendicular to the stairs on the front of Building J intersects an opening to a passage midway up the stairs of Building P. The end of this passage is penetrated by a vertical tube, which may have been intended to permit the zenith transiting sun to shine into the chamber. (Griffith Observatory, after Anthony F. Aveni)

Capella at the front of the building, yet we wonder, is the relationship coincidental?

While we were still pondering the odd shape and situation of Building J, we learned that a similar building existed nearby. About thirty-one miles east of Monte Albán at the top of a small plateau lie the ruins of Caballito Blanco, to our knowledge the location of the only other building in Mesoamerica possessing the same approximate shape as Building J. Our measurements revealed that Building O at Caballito Blanco, constructed about the same time as Building J at Monte Albán, has approximately the same orientation relative to the other structure at Caballito Blanco as does Building J when compared with buildings at Monte Albán; but the absolute orientations are different.

We have no clues regarding the mystery of the peculiar arrow shape of the ground plan at either Monte Albán or Caballito Blanco, but our analysis of Building O once again turned up astronomical directions of significance. The perpendicular to the front of Building O terminated on the elevated northeast horizon exactly at the position of first gleam sunrise on the first day of summer. Here the sun came to rest on its annual migration back and forth along the horizon. Once this point was reached, the sun started its southward progression, which foretold the lengthening of subsequent nights. The priests of Caballito Blanco joined a building and the landscape, thereby creating a solar calendar.

We also checked the direction of the arrow point at Caballito Blanco as had been done for Building J at Monte Albán. This direction reached the southwest horizon, within a ¼ degree of the setting position of Sirius, the brightest star in the sky. Like Capella, this star appears to have been of particular significance at this place and time. The occa-

A second arrow-shaped building, about thirty miles from Monte Albán, is located at Caballito Blanco. (Robin Rector Krupp)

sions of its heliacal rising and setting could have been employed to divide the year into seasons. At Caballito Blanco, Sirius made its heliacal rising close to the summer solstice, and the winter solstice was the last occasion on which it would have been seen rising in the east after sunset.

We can never know for sure whether Building O at Caballito Blanco was actually designed to serve an astronomical function or whether it is a non-functional imitation of Building J at Monte Albán. The two special astronomical relations strongly imply that its builders were again attempting to wed their earthly realm to the heavens, by incorporating significant astronomical directions into their architecture.

THE CARACOL

The Caracol tower at Chichén Itzá in Yucatán is probably the most famous of all the astronomically related buildings in ancient Mesoamerica. So beguiling are its asymmetries that the celebrated American archaeologist J. Eric S. Thompson was moved to write in 1945:

> Every city sooner or later erects some atrocious building that turns the stomach: London has its Albert Hall; New York, its Grant's Tomb; and Harvard, its Memorial Hall. If one can free oneself from the enchantment which antiquity is likely to induce and contemplate this building in all its horror strictly from an aesthetic point of view, one will find that none of these is quite so hideous as the Caracol of Chichén Itzá. It stands like a two-decker wedding cake on the square carton in which it came. Something was pretty clearly wrong with the taste of the architects who built it.

Thompson's frustration with the plan of the tower seems further justified by the American archaeologist Oliver Ricketson's description of the complicated ground plan he encountered when excavating the structure half a century ago:

> The Caracol is a circular tower with four outer doors facing the cardinal points of the compass. Within is a circular corridor from which four more doors, facing midway between the cardinal points, lead into another circular corridor. The inner circular corridor surrounds a masonry core inside which a small spiral staircase leads to the top of the building . . . Near the top of the structure is a flat area from which open three rectangular horizontal shafts. The largest of these, No. 1, faces west and until recently was the only one known. Windows 2 and 3 face southwest and south respectively.

To complicate matters, our measurements on all of the implied symmetries in the foregoing passage suggest the building plan is even more

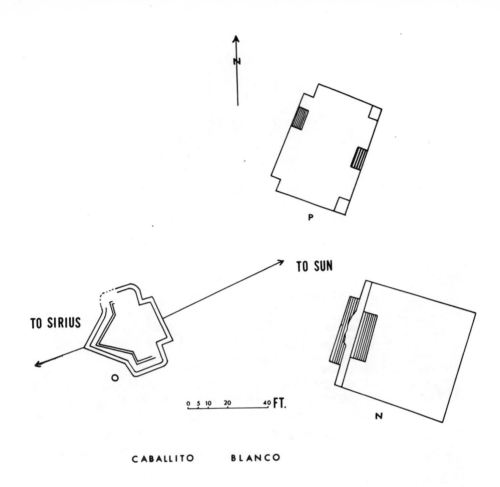

TO SUN

TO SIRIUS

P

O

N

0 5 10 20 40 FT.

CABALLITO BLANCO

The perpendicular to the skewed stairs on the front of Building O at Caballito Blanco points to the summer solstice sunrise. Sirius, which set in the direction of the arrow's point at the left, rose heliacally near the date of the summer solstice. (Anthony F. Aveni)

complex. The doorways are situated 10 to 12 degrees from the cardinal points and the openings in the inner chamber are not symmetrically placed relative to the doorways. The Caracol receives its name from the above-mentioned staircase (more appropriately a narrow passage), which coils upward a full turn, imitating the shell of a snail, or *caracol*.

The Caracol's lack of aesthetic appeal has led some investigators to suggest a functional motivation for its design. It has been called the gnomon of a huge sundial and a military watch tower, but of all its possible uses, astronomical observations seem most successful in accounting for the peculiarities of its structure and orientation.

The Caracol at Chichén Itzá, in Yucatán, included windows in its upper portion that could have been used for astronomical observation. (Robin Rector Krupp)

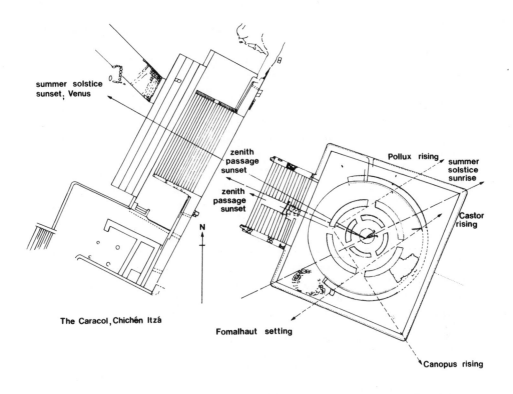

summer solstice
sunset; Venus

zenith
passage
sunset

zenith
passage
sunset

Pollux rising

summer
solstice
sunrise

Castor
rising

N

The Caracol, Chichén Itzá

Fomalhaut setting

Canopus rising

A variety of astronomical alignments appear to be involved in the platforms and lower structure of the Caracol. (Griffith Observatory, after Karl Ruppert and Horst Hartung)

The Lower Platform, the first building unit of the Caracol, was constructed by the Maya about A.D. 800. It is a large rectangular area elevated a few yards above the flat terrain. The front of the building faces 27½ degrees north of west, well out of line with the other buildings at Chichén Itzá. The sunset position at summer solstice lies within 2 degrees, but an even closer match with this direction is provided by the northernmost setting position of the planet Venus, a point on the horizon attained every 2,920 days.

Venus is of special interest in connection with the Caracol, for it is the celestial manifestation of the Toltec god Quetzalcoatl, who was equivalent to the Maya's Kukulcan and is a deity who is symbolized by round structures throughout Mesoamerica. We recall that he is pictured in various evil manifestations in the Venus tables of the Maya *Dresden Codex*, when he makes his first appearance in the predawn sky. The tables give heliacal rise dates and ritually significant dates of appearance and disappearance of the Venus god on the horizon. They contain a means of relating ritual to observations of the planet. Furthermore, the tables were written about the time the Caracol was built, and their point of origin is believed to be not too distant from Chichén Itzá. Were the astronomical observations related to the Venus tables in the *Dresden Codex* executed in this very tower?

Above the Lower Platform, embedded in the stairway of the Upper Platform, we find the Stylobate, a structure consisting of a pair of columns on a small platform aligned asymmetrically relative to the Upper Platform. The columns retain flecks of black and red paint. Since red and black are directional colors in the Maya religion, it is possible that the painted Stylobate could have served as a monument to Venus in the east (red) as morning and in the west (black) as evening star. The base of the platform points exactly to the northern Venus extreme. We find in the writings of Caso a specific association of these colors with Quetzalcoatl-Kukulcan, the Venus god of Central Mexico:

> . . . the flight of Quetzalcoatl from Tula to the mythical Tlillan Tlapal-lan, "the land of the black and the red," and his promise to return from the East in the year of his name, *Ce Ocatl*, is but a mythical explanation of the death of the planet, his descent into the West, where the black and the red, night and day, merge, and the prophecy that he will reappear in the East as the morning star, preceding the sun.

A priest climbing to the top of the tower would have been able to make two more Venus observations in the "windows." Three horizontal shafts look out onto the flat southern and western landscapes. The largest is usually called Window No. 1. It will accommodate a person attempting to crawl through it. The other two are so narrow as to leave no doubt regarding their function. They frame segments of the southern and southwestern horizon.

Ricketson hypothesized that diagonal sight lines, for example, from the inside right to outside left jamb of a window, could have pinpointed quite accurately the position of a horizon event. This hypothesis was given some confirmation when we measured the inside left to outside right alignments in Windows 1 and 2. These directions perfectly framed the northerly and southerly extremes of Venus along the horizon. Furthermore, the sunset on the day of the equinoxes fit neatly into the narrow strip of sky when viewed along the inside right to outside left alignment of Window 1. Since the sunset position on a given date in the seasonal year has scarcely changed in the ten centuries since the Caracol was built, we were able to demonstrate this key alignment by direct photography at the site. Thus, we recorded the setting of the sun in the narrow slot of Window 1 on the first day of spring 1974.

Window 1, the largest opening in the Caracol's upper tower, permits a clear view of the flat western horizon. The height of both the tower and its platforms are clearly advantageous for astronomical observation. (Horst Hartung)

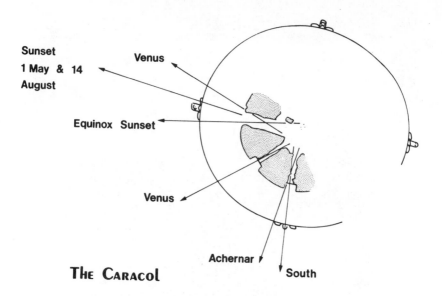

**Sunset
1 May & 14
August**

Venus

Equinox Sunset

Venus

Achernar

South

The Caracol

Only three windows remain in the ruins of the Caracol's upper tower.
(Griffith Observatory, after Karl Ruppert and Anthony F. Aveni)

The sun entered Window 1 about March 20. Sunsets could then be viewed in the window throughout the spring as the sun progressed northward. It passed the midline of the window on April 28. At the solstice "turn around," the sun would have been well to the right of the midline; then it progressed southward, passing out of the window about September 20. Impermanent markers on the window sill could have been used to chart the daily progression of the sun from equinox to equinox, the Caracol functioning as an almanac in stone. Only the three windows at the western side of the building remain today, but the British archaeologist-explorer Alfred Maudslay, who visited Chichén Itzá during the last century, tells us that in the Caracol he viewed,

> an upper storey furnished with what looked like six small doorways facing outwards. Of these, the doorway immediately over the lower doorway . . . is the entrance to a small passage, three feet high, which probably passed right across the building to a doorway on the other side.

Window 1 may have had its counterpart on the other side of the building to follow the solar sunrise path during the winter season.

In the class of possible solar observatories exemplified by Window 1

The vernal equinox sunset (1974) fits perfectly into the narrow wedge of sky visible across a diagonal of Window 1 in the Caracol. (Anthony F. Aveni)

of the Caracol, perhaps one should include an architectural group at Uaxactún in the Petén rain forest of northern Guatemala. Standing on the top of what is known as "Structure E-VII sub," also known as the Pyramid of the Masks, one looks out toward the east over an open plaza, now overrun by jungle. On the other side of the small plaza are three mounds, the outside ones equidistant to the north and south of east from the central mound, which is situated due east. The observer sees the sun rise over the northeastern mound on the summer solstice, over the southeastern mound on the winter solstice, and over the center mound at the equinoxes. Moreover, this plan is duplicated, at least in form if not in orientation, at a dozen more sites within a sixty-mile radius of Uaxactún. A careful reverification of Uaxactún's alignments is needed, however, for the mounds are rather close to the pyramidal backsight. The heights of the original structures on the mounds could have obscured the astronomical alignments.

Another Chichén Itzá sightline worthy of consideration at the top of the Caracol tower is the inside right–outside left diagonal of Window 3. It lies nearly along the present direction of the magnetic compass in northern Yucatán. Since the deviation of the compass needle from astronomical north is strongly time-dependent, it would be dangerous to conclude that ten centuries ago it was the same as it is at present. On the other hand, there is some evidence to suggest that the Mesoamerican people may have known of the concept of a magnetic compass. At San Lorenzo, in Veracruz, Michael Coe discovered a flattened, oblong piece of magnetite with a longitudinal groove along one surface. When suspended by a thread, the bar consistently pointed toward magnetic north. Moreover, many buildings in Yucatán align closely with the same direction.

One must not overlook the possibility that entire constellations or star groups could have been viewed through the Caracol windows. The wide field of view of Window 1 (which exceeds 25 degrees horizontally) supports such a hypothesis. An observer in A.D. 1000 would have seen the sun set along the midline of the window on April 28. Simulation of the view out Window 1 in the planetarium of the University of Florida permitted a quick discovery of what happened shortly after sunset on the same day in that era. A conspicuous group of stars, the Pleiades, sets near the lower right corner of the window. The Pleiades would have disappeared from the window as Aldebaran entered at the left. The Pleiades, as we have seen, have been recognized as functionally one of the most important groups in Mesoamerican star lore, and recent ethnographic accounts confirm that the modern Chorti Maya people of eastern Guatemala are as interested in observation of constellations as their Mesoamerican ancestors were. The Pleiades are still watched carefully on the dates of zenith passage, although precessional changes have elevated other constellations (like Orion's belt) to a calendric significance they may not have possessed in the eleventh century.

Though our investigations throughout Mesoamerica reveal a number of significant astronomical events occurring along the directions of many of the measured alignments, it is clear that not every alignment appears to have an astronomical match which we can recognize. It may be that only some of the sighting possibilities we have discussed were actually functional. Moreover, our search of significant astronomical events to match the alignments has included only those which seem of obvious importance to us: solar, lunar, and planetary extremes and the setting positions of the brightest stars which announce, through their heliacal rising and setting, important dates in the civil, religious, or agricultural calendar. We infer no grand cosmic plan for the architecture of the Caracol tower, but we do imply that the building, apart from being a monument dedicated to Quetzalcoatl-Kukulcan, was erected primarily for the purpose of embodying in its architecture certain significant astronomical directions, in the same sense that a modern almanac exhibits information of importance to us in the keeping of the current calendar.

Other round structures resembling the Caracol exist in the Yucatán peninsula. The ruined tower at Mayapán, in Yucatán, described so aptly by John Lloyd Stephens, was surely a thirteenth-century copy of the Caracol:

> It stood on the ruined mound thirty feet high . . . the building was circular . . . The exterior is of plain stone ten feet high to the top of the lower cornice and fourteen more to the upper one. The door faces *west* and over it is a lintel of stone. The outer wall is five feet thick; the door opens into a *circular passage* three feet wide and in the center is a *cylindrical mass of solid stone* [author's italics].

A similar tower at Paalmul on the eastern coast of the Yucatán peninsula has four different walls of masonry, a west-facing doorway, and a small observing chamber in the uppermost turret. South of Chichén Itzá at the northern edge of the Petén rain forest lies the small tower of Puerto Rico. It is a solid cylinder of masonry pierced only by horizontal orifices barely large enough to accommodate an eye view. The plan of the shafts resembles that of the surviving Caracol windows. Since none of these towers has been investigated for astronomical orientations, it is impossible to know whether they are only imitations of the Caracol. Possibly a system of tower observatories existed throughout Yucatán for accurate calendar keeping on a nationwide basis.

VENUS AND THE GOVERNOR'S PALACE

The Maya expression of direction indicators is quite different at another great Maya city to the west of Chichén Itzá. At Uxmal, most of the structures are oriented 9 degrees east of north. A notable deviant is

the Governor's Palace, which is erected on an elevated artificial platform skewed 20 degrees to the common axis. From its central doorway,
precisely along a perpendicular to the façade, we were able to view a
man-made structure several miles distant. Only a poor untraveled path
made way to these ruins from the only paved road in the vicinity, but
after much difficulty we were able to locate the structure. It stood
among a surprisingly large field of ruined mounds. Stephens, that in-
defatigable explorer, had been there before us and learned that the
name of the city was Nohpat. His description of the ruins and the view
looking back toward Uxmal reads like a fairy tale:

> With the ruins of Nohpat at our feet, we looked out upon a great
> desolate plain, studded with overgrown mounds, of which we took the
> bearings and names as known to the Indians; toward the west by north,
> startling by the grandeur of the buildings and their height above the
> plain, with no decay visible and at this distance seeming perfect as a
> living city, were the ruins of Uxmal. Fronting us was the Casa del Govern-
> ador, apparently so near that we almost looked into its open doors and
> could have distinguished a man moving on the terrace; and yet for the
> first two weeks of our residence at Uxmal, no part of it was visible from
> the terraces or buildings there.

The perpendicular we measured from the Governor's Palace also
points exactly to the southerly rising extreme of Venus. While at
Chichén Itzá the interest in Venus was manifested architecturally in a
circular building, the Maya of Uxmal chose to orient a large rectangular
building to face the planet.

American anthropologists Kent Flannery and Joyce Marcus have
suggested that some Maya centers appear to display a simple geometric
relationship to others nearby. Secondary centers seem to be nearly
equally spaced in a latticelike arrangement around a primary center.
They have suggested a large-scale structural plan with four regional cap-
itals to emphasize the Maya four-directional view of the universe. Each
city is the geometric center of a cluster of secondary centers. Thus,
cosmology may have played a major role in the large-scale territorial organization of the Maya. We do not know whether the relative location
of other centers was determined astronomically but the planned location of Nohpat relative to an important building at Uxmal seems evident.

SOME CONCLUSIONS

In discussing the role of astronomical considerations in Mesoamerican
architecture, we have tried to apply a systematic quantitative approach,
one which has hitherto been seriously lacking in this speculative field.

The ruins of Nohpat, in Yucatán, are visible, as a small mound (indicated by the point of the arrow) on the relatively flat horizon, from the central doorway of the Governor's Palace at Uxmal. This long line-of-sight also includes the phallic stela and altar of the double-headed jaguar just in front of the Governor's Palace and coincides with the most southerly rising of Venus. (E. C. Krupp)

Attempting to look at all the possibilities which make sense, we have stressed astronomical events of obvious functional importance. One of the most puzzling facts about building orientations in Mesoamerica is the east-of-north arrangement of so many ceremonial centers spread throughout the area over such a long time base. A single astronomical explanation is not possible for all that we see. The most likely influence of astronomy upon architecture appears in the misshapen structures like Building J at Monte Albán and the Caracol at Chichén Itzá. But the question of which astronomical objects actually served as orientation points can never be decided on an absolute basis, and it would be inappropriate to apply statistical considerations to the argument. Venus and the Pleiades are frequently excellent matches for some of the alignments; both are mentioned in native Mesoamerican folklore and legend. But Sirius and Capella are not. While our computer tells us that these stars could have served to mark special dates in the calendar, the ethnohistoric sources do not support their importance. Perhaps more attention should be paid to the surviving traditions and current folk practices relating to astronomy among the people of Mexico and Central America.

We have seen in these pages several cases suggesting that astronomical observations were of great importance to the people of ancient Mesoamerica. In view of the fanatical devotion of these people to calendar keeping, which must surely derive from astronomical observations, it seems logical that astronomical baselines would be incorporated into their architecture, especially in Yucatán where the flat, featureless terrain offers no natural foresight for making horizon observations. Objective evidence of the type presented here lies at the foundation of any study of building orientations, but it must be supplemented by the interpretations of a wider community of scholars—archaeologists, ethnologists, historians of science, and anthropologists, all of whom are involved in the interdisciplinary field of archaeoastronomy.

CHAPTER SIX

Astronomers, Pyramids, and Priests

E. C. KRUPP

The few remaining written records of the Maya and others provide a tantalizingly incomplete image of ancient Mesoamerican astronomy. "If only," we find ourselves speculating, "if only a few more codices were known, we might be able to puzzle out the picture." Egypt, by contrast, has left us many hieroglyphic inscriptions—but for all the knowledge we have about ancient Egypt, most Egyptian astronomical lore is lost to us.

Ambiguities in the style of Egyptian expression have precipitated what seem to be interminable arguments about what the Egyptians really had in mind. Many ancient inscriptions and monuments have been turned to literary use by occult and metaphysical societies. Those most visible remains of ancient Egypt, the pyramids, have been particularly abused by ranks of enthusiastic interpreters.

Even when the smoke clears and the genuine accomplishments of the Egyptian astronomers—for example, the tropical calendar and the alignment of the pyramids—are recognized, we still find them relatively unappreciated by historians of science. It may be that the Egyptians were no match for the Babylonians, with their highly systematic observations and their mathematical astronomy. But Egyptian astronomical thought is intertwined with myth and allegory and does represent an interesting phase in the development of culture. It deserves more careful study for that reason. We may yet be surprised by what we find among the tombs and temples of the Nile.

Thirty-three centuries ago a heretic king challenged the religious and political power structure of his land, Egypt, and proclaimed the actual disc of the sun, Aten, to be the one true living god. In doing so, Amenhotep IV dispensed with gods in the guise of the human form. He renamed himself Akhenaten, or "Pleasing to Aten," and, in the words of the hymn he composed himself, exalted the sun:

> Beautiful is your shining forth at heaven's edge,
> O living Aten, beginning of life!
> When you arise on the eastern horizon,
> You fill every land with your beauty.
> You are lovely, great, and glittering
> And go high above every land.
> Though you are far away, your rays are on earth.
> Though you fill men's eyes, your footprints are
> unseen.

Akhenaten moved the capital from Thebes to a new site, Tell-el-Amarna, in order to diminish the power of the priesthood of the orthodox sun god, Amen-Ra. The duration of Aten's supremacy was short, however. It ended with the end of Akhenaten's twenty-year reign as pharaoh. The political strength of the priesthood of Amen-Ra returned the capital of Egypt back to Thebes.

Egyptian religion was rich with solar imagery. Amen-Ra, who was displaced by Aten and who in turn replaced him, was also a manifestation of the sun. In addition to status as chief god of Thebes, Amen-Ra held the power of the sun at its greatest strength, at the zenith.

Amen-Ra was often symbolized by the ram. The Great Temple of Amen-Ra at Karnak included an avenue of ram-headed sphinxes. The presence of the ram suggests more astronomical imagery. It was a symbol of fertility, but it also may have indicated the presence of the vernal equinox in the constellation of Aries, the Ram, during the first and second millennia B.C. So little is known for certain about the Egyptian constellations, however, that it is safer to regard the sun-ram association as evocative but insufficient evidence of an awareness of the "Age of the Ram."

The sun was important in even the earliest Egyptian religion. The tribal cultures in the Nile River valley identified the god Horus, the Falcon, with the sky. The sun was the right eye of Horus and the moon his left. Among the Horus worshipers the sun was not regarded as a beneficent god. It was, instead, a destructive force and an enemy of farmers. Rather it was the Nile that was recognized as the source of cosmic good will. The Nile, with its annual flooding, made civilization possible in Egypt and guaranteed successful agriculture and economy. The sun was respected for its power, but the Nile was the real ruler of Egypt.

Other peoples, sun worshipers from the Mediterranean or the Caucasus, lived side by side with the Horus peoples, and gradually the cultures merged. In fact, Upper and Lower Egypt were first united by sun worshipers, the priests of Ra. A capital was built at On, or Heliopolis, as the Greeks later called it. This site is just a few miles northeast of modern Cairo, and it was an important center for Egypt's solar religion.

We might expect to see a highly developed astronomy in a society like Egypt's, where astronomical imagery was so much a part of its mythology. We do know that the same components of the natural environment that were of such interest to other ancient peoples were of interest to the Egyptians too. Their cosmos was oriented and energized by the sun, moon, and stars. The sky, the earth, the air, and the water were both stage and actors in the drama of the universe. Yet from the evidence that is available, Egyptian astronomy was inferior to its Babylonian counterpart. New evidence may eventually demonstrate greater astronomical sophistication among the ancient Egyptians, but for the present nothing comparable to the systematic, mathematical astronomy practiced between the Tigris and Euphrates rivers is known.

By the Ptolemaic period (332–30 B.C.), Egyptian astronomy bears more of the expected trappings, but by then the influence of many outside forces had been felt. Even then, there was no systematic and comprehensive observation of the sun, moon, planets, and stars, or at least no evidence of such observations remains. Nor is there any distinctly Egyptian technical vocabulary for astronomical phenomena. Despite these shortcomings, Egyptian astronomers must have existed, for their contributions to modern civilization—the tropical calendar and the twenty-four-hour day—continue to remind us of the ideas they developed more than five thousand years ago.

A RIVER AND A STAR

The calendar used in the Western world today is a solar calendar and is based upon the length of the tropical year—the time, say, from one summer solstice to the next. Julius Caesar was responsible for introducing it to Western civilization, and he assigned an Egyptian, Sosigenes of Alexandria, to do it. The Greeks had always relied upon a lunar calendar, and even such expert astronomers as the Babylonians restricted themselves to a calendar based on the moon. But it is very difficult to keep a lunar calendar in step with the natural cycle of the seasons and the solar year; it is typically cumbersome and erratic and unsatisfying to a pragmatic mentality.

Whatever the Egyptians' worth as astronomers, they were known for their reliable solar calendar. It was ideal for civil use and the concerns of empire, and so Julius Caesar had it adapted by Sosigenes for Rome.

In Egypt's earliest days, the protodynastic phase, a lunar calendar was in use. Its months were measured differently from those in Mesopotamia, however. There the month began with the first appearance of the new crescent of the waxing moon, in the west, at sunset. The Egyptians measured their month from the morning when the old crescent of the waning moon became invisible in the predawn eastern sky.

Egyptian astronomers kept their lunar calendar in step with the seasons by intercalating an extra month whenever necessary. Observations of the heliacal, or predawn, rising of Sirius, the brightest star, told them when the extra month was needed.

Heliacal risings are particularly convenient celestial references. Stars rise and set through all hours of the day and night, but, as noted before, if a star, or planet, is barely visible in the eastern sky before sunrise, the star is said to rise heliacally.

Through the course of the year the sun appears to move eastward among the background stars. The change is roughly 50 arc minutes per day. Stars which are east of the sun will rise after sunrise. Therefore they rise in the daytime, and their risings go unseen. As the year progresses, however, the sun gradually moves eastward far enough to permit a particular star to rise with the sun, but the star is still invisible in the glow of dawn. After about twelve more days the sun moves 10 degrees farther east, and the star, on one specific morning, becomes visible to an attentive observer who has been waiting for the star's first reappearance.

By coincidence, during the period of development of the Egyptian calendar, the star Sirius rose heliacally at about the same time of year that the Nile overflowed its banks again to bring life to the farmland on which Egyptian society so greatly depended.

The apparent connection between celestial and terrestrial phenomena greatly affected the Egyptian view of the world. The Egyptians considered the heliacal rising of Sirius to be so important that they marked the beginning of the new year by this event. Even more compelling was the fact that the heliacally rising Sirius and the rising Nile coincided, approximately, with the summer solstice. This most northerly of sunrises did not go unnoticed, and its importance would become even greater when the Egyptians, for administrative purposes, adopted a tropical calendar.

The Egyptian lunar year divided quite naturally into three seasons, each containing four months. The seasons were based upon the behavior of the Nile River and were named Inundation, Planting and Growth, and Harvest and Low Water. The twelve lunar months fell short of a tropical year by eleven days, however, and it was necessary to insert an extra month every two or three years to keep the lunar months and the seasons co-ordinated. The fourth month of the third season, the twelfth and last month of the year, was named "Sirius Rising." The close of the lunar year, coming, as it did, sooner than the end of the solar year, fell closer and closer to the heliacal rising of Sirius. Whenever Sirius was observed to rise during the last eleven days of the month of Sirius Rising, an intercalary month was added. In this way the lunistellar year was resynchronized and its liturgical use continued. Seasonal festivals were determined by its calendar.

GETTING DOWN TO BUSINESS

By early in the third millennium B.C., administrative and fiscal needs had stimulated the Egyptians to introduce a solar calendar based upon the tropical year. The exact length of this year could have been determined by counting the days from one heliacal rising of Sirius to the next.

In contrast with the lunistellar calendar's months, each month in the solar calendar was assigned thirty days. The lunar months were subdivided into four weeks, which in turn were based upon the moon's phases. The thirty days of a solar month could not conveniently be divided by four, and so these months of the civil calendar were each split into three ten-day periods. Rather than start each period with a particular phase of the moon, the Egyptians began each ten day "week" in the civil calendar with the nightfall rising of a particular star or asterism.

With twelve months, each of thirty days duration, and five extra, or epagomenal, days at the end, the Egyptian civil year. totaled 365 days. The true length of the tropical year is closer to 365¼ days. Although from early on the Egyptians knew about the fourth-of-a-day error, they never did anything about it, and it wasn't until the time of Julius Caesar that a sixth epagomenal day was added every fourth year.

Omission of the extra fourth of a day threw the Egyptian civil calendar out of step with the seasons. It is said that the Egyptians were satisfied, however, that after 1,461 of their civil years, their calendar would get back in step. Heliacal risings of Sirius, alleged to complete a cycle of calendar dates in 1,461 years, were responsible for giving this interval the name "Sothic cycle," for Sothis, the Greek name for Sirius. The Egyptians merely noted the existence of this Sothic cycle and let their civil calendar slowly advance through the seasons and through heliacal risings of Sirius, one fourth of a day per year.

Considerable discussion and confusion surrounds the subject of the Sothic cycle. It has often been said that it was 1,460 years long, but that is in terms of Julian years of 365¼ days each. In terms of the Egyptian civil calendar year, or "vague year," of 365 days, the Sothic interval is 1,461 years. It might be thought that the annual precession of Sirius would alter this result, but by chance, the precessional movement nearly cancels the difference between the Julian and the Gregorian years. The Gregorian year is named after Pope Gregory XIII, who in 1582 reformed the calendar of Julius Caesar by permitting only century years divisible by 400—A.D. 1600, for example—and noncentury years divisible by 4 to remain as leap years. This brought the average length of a calendar year closer to the actual time it takes the earth to circuit the sun and so eliminates some of the accumulated error. It has also been pointed out that the relative motion in space between Sirius and the sun is not canceled out, however, and that this makes the Sothic period actually 1,456 Julian years long. Others have also commented on various ambiguities in the Sothic cycle, and Arthur Earle, author of *The Bible Dates Itself*, has proposed, on the basis that heliacal risings could be observed only to the precision of the whole day and not a fraction of a day, that the Egyptian Sothic year averaged 365.25641 days. If this be so, he argues, the Sothic calendar and the vague-year calendar would coincide in cycles that alternate between 1,423 and 1,424 years. Complications like these make interpretation of recorded dates on Egyptian monuments and documents very difficult.

In any case, the Egyptians used a star or a group of stars to signal the beginning of one of their ten-day "weeks." Such a star or group of stars is called a "decan." Just as the Egyptian civil calendar evolved into our present calendar, the decans generated a system of timekeeping that led to our use of a twenty-four-hour day, as we shall see shortly.

STARS OVER EGYPT

The use of the decans was partly deduced from analysis of Egyptian star clocks. At some early and unknown point in Egyptian history the star clocks were devised and by 2400 B.C. were definitely in use. The star clock is not a mechanical contrivance but a diagram, and it indicates the name of each decan that is rising on a particular date and at a particular time, or hour, of night. A sky observer could use the star clock, by knowing the date, to calculate the hour of night.

All twelve known examples of star clocks are found on the inside of coffin lids. They date from the Ninth to the Twelfth Dynasties (2160–1786 B.C.). Inclusion of the clock into the paraphernalia of the dead indicates how vital the star clock was to the Egyptians. It obviously was regarded as a necessity in the afterlife.

A star clock is simply a grid, and each square represents a date and a time. The decan named in each space is that whose rising coincides with that time. The diagram is read from right to left, and the first thirty-six of the forty columns represent the thirty-six ten-day periods, or decades, in the Egyptian civil year. The next three columns are extraneous and simply list thirty-six stars, or decans, twelve in each column. The fortieth and last column represents the epagomenal days.

The twelve rows represent the hours of the night. The decision to subdivide the night into twelve intervals derived from the significance attached by the Egyptians to the heliacal rising of Sirius. To understand this connection we must analyze the sequence of decans in the star clock.

If a star is observed to rise heliacally, it might be said to mark the night's last hour, for sunrise is not far behind. As the days pass, however, the sun appears to move farther eastward on the ecliptic and the star that rose heliacally becomes visible above the horizon earlier each morning. Ten days after the first heliacal rising it is necessary to find another heliacally rising star to mark the night's last hour. The previous star now marks the next to the last hour.

The bottom row in the star clock represents the last hour of the night. The decan named in this row and in the first column on the right is the heliacally rising star for the first decade of the year. One decade later this star should be named in the second row from the bottom (next-to-the-last hour of the night) and in the second column from the right. This is exactly the diagonal pattern found in the star clocks. Any particular decan was used to note the twelve successive hours of the night over a period of 120 days, or twelve decades.

It seems at first curious that the Egyptians chose to divide the night into twelve parts, for one might guess that these "hours" would instead equal the number of decans which rise between dusk and dawn. With thirty-six decades in the year and thirty-six decans around the sky, eighteen, or half of the decans, might be expected to rise during the night. Why, therefore, didn't the Egyptians divide the night into eighteen parts?

The duration of darkness varies throughout the year. The Egyptians

Epagomenal Days | **DECADES**

40	39	38	37	36	35	34	33	32	31	30	29	28	27	26	25	24	23	22	21	20	19	18	17	16	15	14	13	12	11	10	9	8	7	6	5	4	3	2	1	hours of the night
A 25	13	1																										12											1	I
B 26	14	2	A																										12										2	II
C 27	15	3		A																										12									3	III
D 28	16	4			A																										12								4	IV
E 29	17	5				A																										12							5	V
F 30	18	6					A																										12						6	VI
G 31	19	7						A																										12					7	VII
H 32	20	8							A																										12				8	VIII
J 33	21	9								A																										12			9	IX
K 34	22	10									A																										12		10	X
L 35	23	11										A																										12	11	XI
M 36	24	12	L	K	J	H	G	F	E	D	C	B	A	36	35	34	33	32	31	30	29	28	27	26	25	24	23	22	21	20	19	18	17	16	15	14	13	12	XII

1— 36   **Regular Decans**

A— M   **Epagomenal Decans**

EGYPTIAN COFFIN LID STAR CLOCK

The shifting rising time of a decan is visible as a diagonal pattern through the ten-day periods, or decades, of the Egyptian calendar. This diagonal format is characteristic of star clocks painted on coffin lids. (Griffith Observatory, after R. A. Parker)

structured their calendar and timekeeping by the behavior of Sirius and the duration of night at the time of its heliacal rising. In ancient Egypt this occurred near the summer solstice, the time of shortest nights, about six modern hours long, and when only twelve decans can be observed. The hours of night were therefore determined to be twelve as a direct and natural consequence of the organization of the Egyptian civil calendar.

Because the duration of night is not constant throughout the year, the Egyptians allowed the hours to vary in length. The pattern of Sirius at the summer solstice became the plan for the entire year, for all of the decans were chosen from stars located in a band south of and parallel to the ecliptic.

A case has been argued by Cyril Fagan that the Egyptians really divided the year into seventy-two "pentades" and that the stars that have been recognized as decans really rose heliacally in five-day intervals. Fagan claims that some of these stars were deliberately omitted from the coffin-lid star clocks to avoid reference to seasons other than spring, the season which prevailed in the afterlife. Although some of Fagan's astronomical arguments and identifications are compelling, his unorthodox view is inconsistent with the explanation of the twenty-four-hour day. Most Egyptologists remain convinced that the decans behaved in accordance with their name and rose heliacally in ten-day intervals.

The Egyptians originally divided the daytime, from sunrise to sunset, decimally, but eventually they added two hours, one for morning twilight and one for evening. Twelve variable "hours" of daylight and twelve variable "hours" of night by the twelfth century B.C. finally evolved into a total daily cycle of twenty-four equal intervals of time.

Shadow clocks, and later water clocks, measured the passage of time for the Egyptians in the daytime. The shadow clocks were small, portable instruments, but by Roman times it was thought that large, permanently placed obelisks, like the so-called Cleopatra's Needles which now stand on the Thames Victoria Embankment and in New York's Central Park, had been used for timekeeping. The Romans re-erected an Egyptian obelisk for this purpose in Rome. The famous obelisk in London was actually first erected in 1450 B.C. by Thutmose III and had nothing to do with Cleopatra. Brought to London in 1877, it had in fact been given in 1820 to George IV, on his accession to the throne, by Mehemet Ali. Another obelisk, also the gift of Mehemet Ali, had been re-erected in the Place de la Concorde, in Paris, in 1836.

Obelisks were monuments to Amen-Ra. The pyramidion at the top was usually plated with metal—copper, gold, or electrum—and an association with the sun seems clear. How the ancient Egyptians used them for timekeeping, if they did, is ambiguous.

Procedures of Egyptian timekeeping were modified as the centuries passed. The quarter-day error in the civil calendar necessitated revision of the decans, and by the Twelfth Dynasty, 2000 B.C., decan transits, not risings, were observed. By about 1500 B.C. diagonal star clocks had been abandoned, and a new type of star clock, the Ramesside star clock, was invented.

The Ramesside star clock required observation of decanal stars in transit and on either side of the meridian. The specified stars were sighted with merkhet (plumb line) and bay (forked stick). The hour on a particular date was given by a decan's position relative to the body of one member of a pair of astronomers. The two astronomers sat facing each other on a north–south line, and the northernmost, the observer, watched the progress of stellar transits in the sky behind his partner to the south. The observer apparently obtained a reading by superimposing his plumb line upon a star. For meridian transit timekeeping, the partner's plumb line (which he held over his own head), the observer's plumb line, the slit on the observer's bay, and the line imagined to lead from the star straight to the ground, all had to be in alignment. A known star would satisfy this condition at a known hour in a

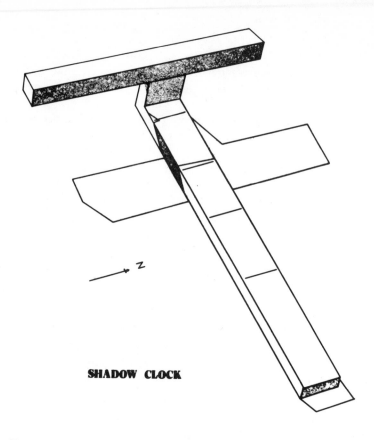

SHADOW CLOCK

The Egyptians could not use stars to measure the passage of time in the daytime, but they did use simple shadow clocks for this purpose until more sophisticated water clocks came into use. A shadow clock was portable and consisted of a main staff a couple of feet long, with a smaller raised crossbar at the end of the main staff. The long end of the staff was pointed toward west in the morning so that the crossbar was on the east. The morning sun, as it rose higher and higher toward noon, cast a shadow of the crossbar on the main staff. The main staff was inscribed with marks to indicate the hours of the day. These were spaced at varying length from each other as a device to tell the hour when the shadow crossed the mark. In the afternoon, the shadow clock was turned around so that the crossbar was on the west side, and the entire process was reversed. After the sun went down, the shadow clock was useless, but then the stars started to become visible. (Griffith Observatory)

Merkhet (plumb line) and bay (forked stick), now in the Kensington Science Museum in London, were aligned on stars. (E. C. Krupp)

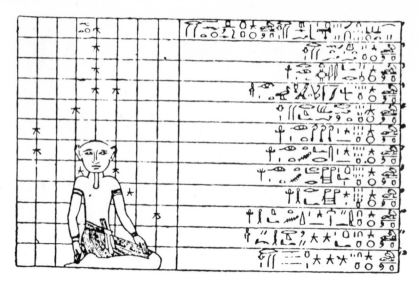

Ramesside Star Clock

The positions of the decans relative to the body of an astronomer's assistant were noted in the Ramesside star clocks. (Moses B. Cotsworth)

particular ten-day period. The Ramesside star clock itself is just a list of the names of the decans and a drawing of the southern member of the pair of timekeepers, with the stars' positions shown around him.

STARS ON THE CEILING

Most of what is known of Egyptian astronomical lore is found on monuments, and eighty-one monuments are known to have some astronomical content. Most of these are tomb and temple ceilings.

The earliest known astronomical ceiling is that of the Tomb of Senmut (1473 B.C.). (Senmut was the architect who built Queen Hatshepsut's magnificent mortuary temple at Deir el-Bahri. Two tombs for Senmut have been found and the one with the astronomical ceiling [Egyptian tomb 353] is unfinished and located below a quarry just east of Hatshepsut's temple.) A star clock, accompanied by representations of planets and certain constellations, is incorporated in the ceiling of this tomb. Isis and Osiris are depicted in boats and represent Sirius and the constellation of Orion, respectively. Jupiter and Saturn are also shown, as falcon-headed gods. Mercury is represented by the god Seth, the enemy of Osiris. Venus appears as a heron. Two turtles stand for one of the decans, and Mars, on this ceiling, is omitted.

On the ceiling's northern half a row of deities of the lunar month stands next to Isis. She, in turn, is accompanied by a set of constellations known to Egyptologists as the "Northern Group." This same group is also included on the ceiling of the tomb of the pharoah Seti I, and some of the Egyptian constellations can be identified with those we use today.

Stars of the Big Dipper were included by the Egyptians within a region of sky symbolized by a bull standing on a platform. Sometimes the same region was represented by just the bull's foreleg.

One of the most engaging Egyptian constellations is an erect hippopotamus. A crocodile is on its back, and its forefeet rest upon a mooring post that is associated with the celestial pole. The hippopotamus includes stars of Draco, the Dragon.

Unfortunately, it is not known how early the Egyptians discovered and named the planets visible to the unaided eye, but planets are represented on the astronomical ceilings of some tombs and temples, as already described. The superior planets visible to the unaided eye were all considered to be aspects of Horus, the sky god, and were named accordingly:

>Saturn =Horus, Bull of the Sky
>Jupiter=Horus Who Illuminates the Two Lands
>Mars =Horus of the Horizon, Horus the Red

All three planets were always depicted with the falcon head of Horus and were usually separated from Mercury and Venus, the inferior planets. Mercury was known both as Sbg, a name whose meaning is lost, and as Seth, the villain in the Osirian cycle. By 1150 B.C. Mercury was known to be the same object in its morning and evening star aspects. As an evening star Mercury took on the malevolent character of Seth, but whether the planet conjured up the same association as a morning star is unknown.

Venus was called "the Crosser," or "the Star Which Crosses," and was pictured with the head of what may have been a heron, or possibly bennu, a heronlike bird. Later Venus too was given the falcon head of Horus, and still later it was shown sometimes with two falcon heads or as a two-faced falcon. The name for Venus and the curious doubling of its head suggest that this planet, like Mercury, was recognized as both a morning and evening star.

THE CEILING OF HATHOR FROM DENDERA

The most spectacular representation of the Egyptian sky is the ceiling from the Temple of Hathor at Dendera in Upper Egypt, the famous circular Dendera zodiac, now in the Louvre. All available evidence indicates that the concept of the zodiac was not native to Egypt, however. Instead, it is believed that the zodiac was imported to Egypt from Mesopotamia at some late but unknown date. The earliest known representation of the zodiac in Egypt was found at the now-destroyed temple at Esna, also in Upper Egypt, which cannot be dated earlier than 246 B.C. and was perhaps nearly seventy years more recent than that.

At Dendera the temple has been dated to 30 B.C. and is, therefore, a relic of Ptolemaic times, as is the circular zodiac. Sir Norman Lockyer, the British astronomer who at the end of the nineteenth century investigated the astronomical potential of Stonehenge and other Megalithic monuments, also visited Egypt and examined temples there for evidence of astronomical alignment. He maintained that the present temple at Dendera is located on a site of much greater antiquity and, further, that the Temple of Isis at Dendera incorporated an alignment with the star Sirius. This alignment dated the temple to 700 B.C. Fortuitously, the nineteenth-century scientist Jean Baptiste Biot had claimed that his analysis of the Dendera ceiling also implied a date of 700 B.C. Biot identified a falcon perched on a papyrus pole with Sirius, and his astronomically determined date depended upon this interpretation. It was subsequently proved that Biot misidentified this and other stars and constellations. Although Lockyer bolstered his argument for Dendera's antiquity with a reference to a Fourth Dynasty text that referred in turn to a stellar alignment of the Temple of Hathor at Dendera, his results have not been checked nor corroborated in more recent times.

No matter what date is assigned to the Dendera circular zodiac, it is still a very elegant example of the Egyptian view of the heavens, for many identifiable constellations and planets are shown on it. The familiar northern constellations of the Bull's Foreleg (Ursa Major, the Big Dipper) and the Hippopotamus (Draco, the Dragon) are accompanied by a properly offset ring of zodiac constellations.

Pisces is shown as two fishes, joined by a cord and accompanied by a wavy symbol that suggests the aquatic attributes of this constellation. Ascending counterclockwise we encounter Aquarius, the Water Carrier, who is pouring two wavy streams of water into the mouth of Piscis Australis, the Southern Fish. Capricornus, the Sea Goat, is next and is followed by a centaur-archer, Sagittarius. A scorpion, Scorpius, leads the archer and is in turn led by Libra, the Scales. The next zodiacal constellation is directly across from Pisces. It is a woman who holds a spike of wheat, an obvious representation of Virgo. Leo, the Lion, follows, and

The ceiling from the Temple of Hathor at Dendera, in Upper Egypt, is now in the Louvre, in Paris. The disc is carved as a constellation map and the zodiac constellations are easily recognized. (Lehner and Lehner)

slightly above him is the Scarab Beetle, which in Egypt took the place of Cancer, the Crab. A man and a woman with joined hands represents Gemini, the Twins. Obvious illustrations of Taurus, the Bull, and Aries, the Ram, complete the set.

Below Leo is situated a cow in a boat with a star between its horns. This figure is Sirius. The jackal near the Hippopotamus is Ursa Minor. Orion is found to be holding a staff and standing near Taurus.

Planets are also on the Dendera ceiling, and their symbols are located in the constellations in which they were thought to be particularly influential, that is, "in exaltation":

> Mercury, in Virgo
> Venus, in Pisces
> Mars, in Capricornus
> Jupiter, in Cancer
> Saturn, in Libra

The disc situated between Pisces and Aries may be the full moon.

Around the perimeter are thirty-six named figures which represent the decans. Other figures represent constellations not yet identified with those in present use.

The Dendera circular zodiac was discovered in 1799 by Napoleon's General Louis Desaix. Its checkered and picaresque history included an attempt to dynamite it from its emplacement in the temple, its theft, and its eventual sale to Louis XVIII. It is now on display at the Louvre in Paris.

Controversy still surrounds interpretation of the circular star map of Dendera. In one unorthodox view, a mark on the breast of the Hippopotamus was identified as the north ecliptic pole. This point and other curious hieroglyphics at the edge of the disc allegedly indicate positions of the equinoxes at a time preceding the carving of the Dendera ceiling. This claim would suggest that the Egyptians were not only aware of the effects of precession but had also deduced the pattern of movement of the equinoxes and north celestial pole through the stars. In *Hamlet's Mill* Giorgio De Santillana and Hertha von Dechend argued that precession has been known and charted since earliest times, that precession was thought to be the fundamental operator of the cosmos, and that precession was thought by the ancients to control celestial and terrestrial activity alike.

Circumstantial evidence implies that the awareness of the shifting equinoxes may be of considerable antiquity, for we find, in Egypt at least, a succession of cults whose iconography and interest focus on duality, the bull, and the ram at appropriate periods for Gemini, Taurus, and Aries in the precessional cycle of the equinoxes.

Comprehensive knowledge of precession seems to be incompatible with the descriptive non-mathematical picture of astronomy that is the natural conclusion of Otto Neugebauer's and R. A. Parker's meticulous analyses of Egyptian astronomical texts. Yet the calendar and the system of decans argue for a certain level of sophistication and observational expertise. In comparison with Babylonian mathematical astronomy, the Egyptian achievement may seem pale, but what there is of it is interesting in terms of the development of human thought. Some believe that the Egyptians expressed precise astronomical information and concepts allegorically, and this may be partially so. Unfortunately, such a circumstance, or even the semblance of such a circumstance, often encourages a variety of untestable hypotheses. This in turn makes it next to impossible to discriminate the correct interpretation from the others.

Whether the Egyptians were fully aware of precession is one thing; that they responded to its effects is another. In Lockyer's view, entire temples were rebuilt to re-establish alignments on important stars.

PASSAGE FOR THE SUN

Sir Norman Lockyer's interpretation of the orientation of Egyptian temples was published in *The Dawn of Astronomy* in 1894, and it created another controversial chapter in the history of archaeoastronomy. Lockyer regularly visited Egypt and investigated temple alignments. He concluded that the Egyptians oriented their temples astronomically and used them as observatories and as ceremonial centers for significant celestial events.

According to Lockyer, the Egyptian temple was designed around a long passage. The temple's long axis was arranged to permit a beam of light from the sun or other celestial object to reach all the way down the passage to a darkened sanctuary at the end. There the light might dimly illuminate the chamber on some appropriate date, for example, the summer solstice.

A very long gallery, perhaps 500 yards in length, would be required to restrict all but the most exactly aligned beams from the sanctuary. The path would be narrowed so that the chamber would be illuminated for only a few moments on just the right day and no others. Lockyer calculated that a properly designed temple would have allowed the Egyptians to estimate the length of the tropical year to one minute's precision, or one ten-thousandth of a year.

Six solstitial alignments were proposed in Lockyer's *The Dawn of Astronomy*, and three of them are at Karnak. One of these is the spectacular Great Temple of Amen-Ra, whose alignment, Lockyer claimed, was toward the summer solstice sunset. He further claimed that one of the other Karnak temples also was oriented on the summer solstice sunset, and a third faced the winter solstice sunrise. This third temple and the Great Temple of Amen-Ra appear to have been constructed back to back.

Of the three other solstitial temples, two may have been oriented to-

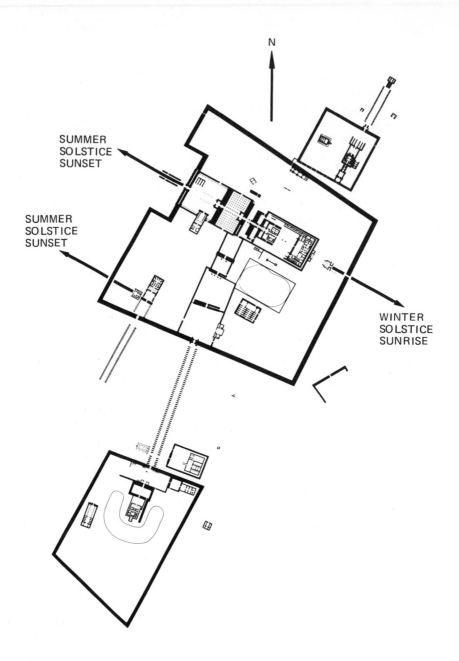

N

SUMMER
SOLSTICE
SUNSET

SUMMER
SOLSTICE
SUNSET

WINTER
SOLSTICE
SUNRISE

Sir Norman Lockyer identified three solstitial alignments at Karnak, in Upper Egypt. Two indicated the summer solstice sunset, he claimed, and the third pointed toward the winter solstice sunrise. (Griffith Observatory, Joseph Bieniasz, after Sir Norman Lockyer) ward the winter solstice sunrise, and the third was, perhaps, set toward the winter solstice sunset.

Lockyer also examined ruins of great antiquity at Heliopolis and Abydos and found less certain indicators of solstitial alignment. On the western side of the Nile, opposite Karnak, two one-thousand-ton statues known as the Colossi of Memnon face the southeast and, in Lockyer's view, the winter solstice sunrise.

Lockyer's detailed discussion of the apertures and pylons of the Great Temple of Amen-Ra at Karnak strongly reinforced his solstitial interpretation of the temple's axis. For other temples which did not follow the solstitial pattern, Lockyer evolved a theory of stellar alignments.

THE TEMPLE AND THE STARS

Stellar alignments of temples, the second fruit of Lockyer's Egyptian studies, involved the assignment of a particular star to a particular temple. This is hazardous unless a reliable construction date for the temple is known. Lockyer proposed a series of solstitially heliacally rising stars that could have been used at different temples and at different times to herald the summer solstice sunrise and the inundation of the Nile. Although such a set of stars might work for this purpose over several thousands of years, the fact that it might is not equivalent to proof that the stars were used in this way.

Despite the difficulty Lockyer faced in proving the existence of stellar alignments, he cited textual evidence from the Temple of Hathor at Dendera and from the temple at Edfu which supported the idea of stellar alignment. In both cases the inscriptions refer to a ceremony of "the stretching of the cord." This ritual was associated with the establishment of the temple, for the temple's axis was laid out by stretching a cord between two stakes. At Dendera the inscriptions indicate that the king had his eye on "ak" in the constellation of the Bull's Foreleg (Big Dipper). Here "ak" may refer to a particular star. A similar reference to the Bull's Foreleg is found at the temple of Edfu, but no particular star seems to be mentioned. Finally, the accounts of the laying of the foundations of these temples also mention the merkhet, and the inclusion of this device, which is known to have been used in connection with stellar observations, further suggests that Egyptian temples were in some way related to certain stars.

Although the temples at Dendera and the temple at Edfu are Graeco-Roman and relatively late (Ptolemaic), I. E. S. Edwards, the foremost scholar on the archaeology of the pyramids, reasoned, in *The Pyramids of Egypt,* that "the stretching of the cord" inscriptions probably preserve a very ancient tradition. He mentioned, in this respect, a relief found in the pharaoh Niuserrē's Fifth Dynasty (2560–2420 B.C.) sun temple at Abu Gurab, near Heliopolis. It shows the pharaoh and the goddess Seshat each holding mallets and stakes, just as they are described at Edfu, for "the stretching of the cord." Lockyer reproduced a similar picture from Abydos and the time of the New Kingdom pharaoh Seti I (1380 B.C.). Yet another confirmation of this surveying technique was discovered at Luxor, adjoining Karnak, where orientation lines were found inscribed on the subflooring in the Sanctuary of the Sacred Boat.

An even more curious hint that Egyptians aligned temples on stars involves precession, the 26,000-year cycle of wobbling of the earth's polar axis. Precession alters azimuths and the rising times of stars in any given season. A heliacal signal of the solstice would only work for a few centuries, therefore, and a temple aligned precisely on a particular star would become useless. Probably a new star would be chosen to herald the opening of the year, and temples would have to be abandoned or rebuilt to a new, correct alignment. In fact, such reconstructions occur. Lockyer noted four additions to the temple at Luxor, each of which deviates slightly from the previous orientation, perhaps to accommodate observation of the star Vega.

An even more striking example is found at Medînet-Habu, in Upper Egypt across the Nile from Luxor. Two temples are located side by side there, but their axes deviate from each other by several degrees. The older is oriented approximately 51½ degrees south of east. It is the smaller of the two and has outside courts as well as a sanctuary. Ramses III built the second, larger temple at Medînet-Habu, and it was oriented 5 degrees farther north. Lockyer calculated that both temples were intended to align with the star Phact, or Alpha Columbae.

Even more significant is Lockyer's conclusion that the heliacal rising of Phact was used to signal the coming summer solstice in the epoch before Sirius could have been used for this purpose, beginning perhaps at some point late in the fourth millennium B.C. Lockyer assembled a set of stars, including Phact, Sirius, Canopus, Capella, Spica, Gamma Draconis, Alpha Centauri, and others on which temples were aligned. In some cases the heliacal risings and settings of these stars were associated with certain dates in the tropical year. Furthermore, at a single temple one star might eventually supercede another as precession slowly carried the former out of alignment.

Although it is possible that the temples were used as Lockyer described, both observationally and ceremonially, his system of alignments is difficult to confirm in detail. His arguments are incomplete, and until a comprehensive picture of practical Egyptian astronomy is

It is known that Egyptian temples were properly aligned at a ceremony called "the stretching of the cord." Astronomical objects were used to establish the reference lines. Here the king and goddess grasp mallets and stakes, around which the cord is looped, and lay the foundation stone. (Sir Norman Lockyer)

available, in which the decan names can be identified with actual stars and which is fully documented by a consistent interpretation of all pertinent inscriptions, the case will remain unresolved. Lockyer's speculative, and in some cases wrong, remarks about Egyptian history and culture led many prominent Egyptologists to dismiss his work. This is unfortunate, for even if in detail his hypotheses are proved wrong, it already seems likely that his approach toward the Egyptian antiquities and his evaluation of Egyptian astronomical endeavor are destined for a rebirth. This is particularly evident in recent work by Gerald Hawkins.

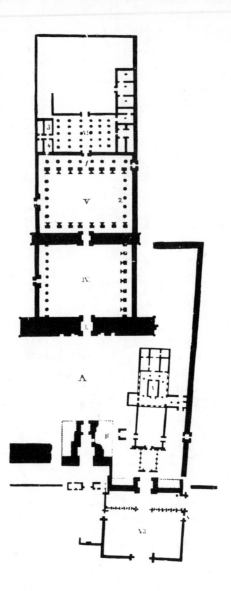

Lockyer interpreted the reorientation of the later buildings (upper) at Medinet-Habu, on the Nile's west bank, across from Luxor, as an Egyptian response to the precession of the equinoxes. (Sir Norman Lockyer)

BEYOND STONEHENGE

Gerald Hawkins, who followed Lockyer's lead at Stonehenge, again traced Lockyer's steps, this time to Egypt and Karnak. In 1973 Hawkins began to publish some of the results of his review of possible astronomical alignments of the Great Temple of Amen-Ra and its associated structures.

In apparent contrast with Lockyer's results, Hawkins concluded that the Great Temple of Amen-Ra was intended to align with winter solstice sunrise and not to the northwest and the summer solstice sunset. According to Hawkins, P. J. G. Wakefield, a British Army engineer, attempted to observe the summer solstice sunset along the temple's axis in 1891 but failed to do so because the western Theban hills elevated the horizon altitude and hid the setting sun. Although most Egyptian temples face the Nile, Karnak is located in the bend of the river and in the only region where solstitial and fluvial alignments might be simultaneously possible. Furthermore, Hawkins cited French Egyptologist P. Barguet's 1962 analysis of architecture and inscriptions at Karnak. Barguet's work suggested that the temple was aligned astronomically away from the river. Without surveying the site himself, Hawkins calculated the southeastern orientation of the temple from old air survey charts and found that it agreed with the 1480 B.C. declination of the winter solstice sun, rising with full disc tangent to the horizon, to ±0.2 degrees. The temple was rebuilt by Thutmose III in 1480 B.C.

It appeared that there was substance to Hawkins' argument, and support for his interpretation comes from several lines of thought. The focus of Egyptian astronomy is known to have been on the eastern horizon. The decans were observed there. Heliacal risings, like that of Sirius, announced important calendar dates. The new month began with the rising last crescent of the moon. The Pyramid Texts refer repeatedly to the morning star and ignore the evening star. Beyond the southeast end of the Great Temple of Amen-Ra, is a temple dedicated to Ra-Hor-Ahkty, the "Brilliant, Horizon Rising Sun," and this temple is apparently open to the winter solstice sunrise.

Although the Great Temple of Amen-Ra appeared to be oriented toward the winter solstice sunrise, Hawkins was troubled by the fact that the temple did not open in that direction. Indeed, the inner sanctum and Hall of Festivals, at the temple's southeast end, block any view of the southeast horizon. The long passage of the temple clearly opens to the northwest.

At the far southeast end of the temple Hawkins examined a small upstairs temple, which he subsequently called the High Room of the Sun, in the building's easternmost corner. A narrow stairway leads up to it,

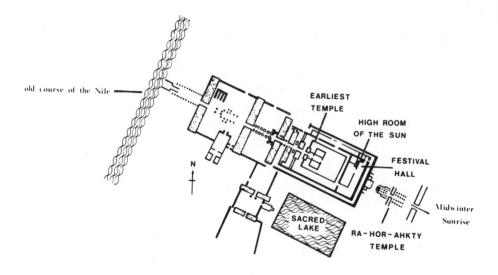

old course of the Nile

EARLIEST
TEMPLE

HIGH ROOM
OF THE SUN

FESTIVAL
HALL

Midwinter
Sunrise

N

SACRED
LAKE

RA-HOR-AHKTY
TEMPLE

Gerald Hawkins concluded that Lockyer had deduced the astronomical alignment of the Great Temple of Amen-Ra at Karnak incorrectly. In the Hawkins version, the structure is oriented to the winter solstice sunrise. (Griffith Observatory, after Gerald S. Hawkins)

and on the wall in the hall below is an inscription that directs the following: "One climbs, the *aha*, the lonesome place of the majestic soul, the high room of the spirit which traverses the sky: one there opens the doors of the horizon of the primordial god of the two countries to see the mystery of Horus shining."

In the upstairs roof temple, which is dedicated to Ra-Hor-Ahkty, one finds an altar and the ruins of a window which opens to the southeast. On the wall the pharaoh is shown facing the window and perhaps greeting the rising sun. A recent wall now blocks the view of sunrise, but the horizon could have been clear in ancient times. Below the temple is another example of a wall mural which illustrates the ceremony of "the stretching of the cord."

Hawkins appears to have found an interesting astronomical interpretation of the Great Temple of Amen-Ra, but in fairness to Lockyer it should be remembered that he did note the winter solstice sunrise alignment of the Temple of Ra-Hor-Ahkty (Lockyer's Karnak Temple

Hawkins, like Lockyer before him, noted that the axis of the Temple of Ra-Hor-Ahkty, located just beyond the southeast end of the Great Temple of Amen-Ra, is oriented to the winter solstice sunrise. Here the sun is framed by the doorway of the temple as it appeared a few minutes after first gleam of sunrise on the winter solstice of 1976, as seen from a sanctuary in the rear, perhaps intended to be illuminated by the southernmost rising sun. (R. L. Mikesell)

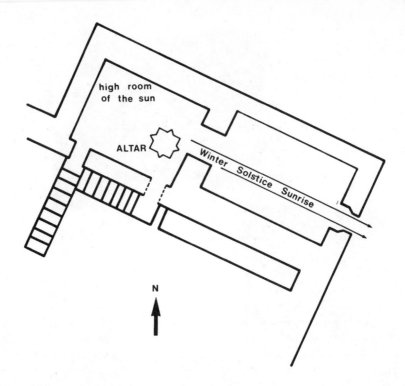

Hawkins found a window in the High Room of the Sun, in the Great Temple at Karnak, which permitted a view to the southeastern horizon. (Griffith Observatory, after Gerald S. Hawkins)

O). Lockyer's conclusion that the Colossi of Memnon face the winter solstice sunrise should also be recalled. The two huge statues once flanked the entrance of a temple whose axis was probably aligned in this same direction. Hawkins confirmed this alignment of the Colossi and found still another alignment at Abu Simbel. At Abu Simbel a small side chapel faces the winter solstice sunrise. The axis of the main temple is aligned with a solar declination of —9.6 degrees. This corresponds to the modern calendar date of October 18 or February 22. The October sunrise in the time of Ramses II agreed with the moveable start of the civil year (1 Peret I), and the whole structure may have been hewn from the rock to mark this significant sunrise at the start of Ramses' thirtieth year jubilee.

Light from the winter solstice of 1976 rising sun penetrates the ruined window frame of the High Room of the Sun at the Great Temple of Amen-Ra at Karnak. (Anna Mikesell)

Several other temples at Karnak are closely associated with the Great Temple of Amen-Ra and are connected to it by avenues. Hawkins has also examined two of these, the Temple of Khonsu and the Temple of Mut, and has proposed lunar alignments for their transverse axes. Khonsu was the moon god and was frequently shown wearing the disc and crescent of the moon on his head. His name means "to travel" or "to travel through a marsh" and is suggestive of the moon's rapid and complex movement through the stars.

Mut was the consort of Amen-Ra, and one of the temples in the northeast part of her complex at Karnak was devoted to Khonspekherod, or "Khonsu the Child." The axis of this temple is perpendicular to the main axis of the main Temple of Mut. Another small temple, built by Ramses III, is southwest of the main temple and parallel to it.

The transverse axes of the Temple of Khonsu and the Temple of Mut may have indicated the tangent setting of the first crescent moon after the summer solstice. The Khonsu line gives the declination for this event at major standstill, and the Mut line provides the minor standstill position 9.3 years later. Even more suggestive is the fact that these major and minor standstill lines, when extended across the Nile to the west bank, pass through the regions of the Tombs of the Kings and the Tombs of the Queens, respectively. If the setting moon, or indeed any setting object, were associated with death, it might be appropriate to place the tombs west of the Nile. Nearly all of the pyramids were built on the Nile's west bank.

If the lunar alignments at Karnak are correctly interpreted, they are unique, for they represent the first lunar alignments known for any Egyptian temple and the first architectural embodiment of the standstill cycle. Although Hawkins reported the declination for the transverse axis of the Temple of Mut, he implied that the perpendicularly arranged axis of the Temple of Khonspekherod was the true lunar alignment, and an association with the new first crescent moon could be implied by the temple's name. It is difficult at first to see how transverse temple alignments would be considered significant in the context of the longitudinally oriented solstitial alignment of the Great Temple of Amen-Ra, but a roof temple with a window open to the southeast appears to duplicate on the Temple of Khonsu the astronomical approach of the High Room of the Sun in the Great Temple of Amen-Ra. Perhaps the transverse axis of the Temple of Khonsu was built with this alignment in mind, but it is even harder to understand why the moon should figure longitudinally with one temple and transversely with another. More serious is the strong case that has been made for the eastern horizon in Egyptian astronomy. In those terms the motivation for commemorating the moonset is unclear.

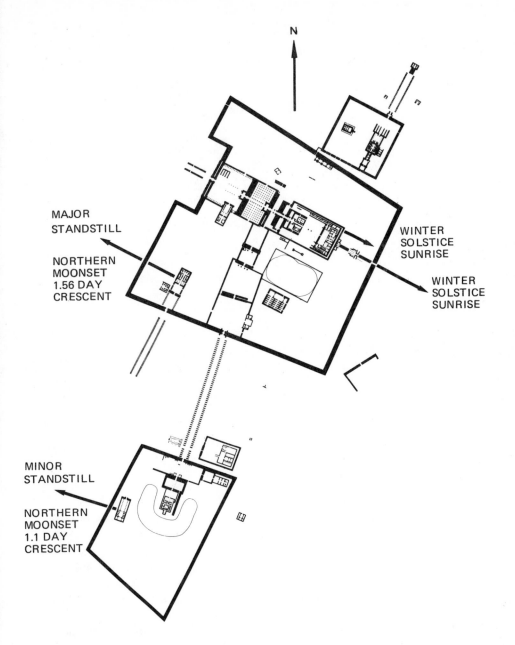

MAJOR
STANDSTILL

NORTHERN
MOONSET
1.56 DAY
CRESCENT

WINTER
SOLSTICE
SUNRISE

WINTER
SOLSTICE
SUNRISE

MINOR
STANDSTILL

NORTHERN
MOONSET
1.1 DAY
CRESCENT

N

Two lunar alignments were identified by Hawkins at Karnak. (Griffith Observatory, Joseph Bieniasz)

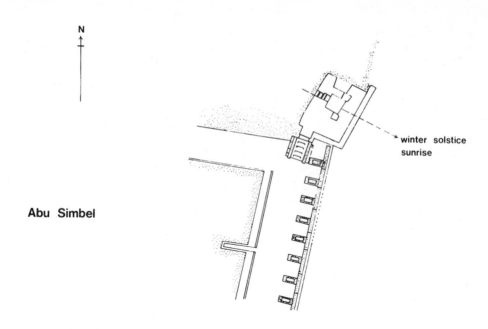

N

Abu Simbel

winter solstice
sunrise

*The main temple of Abu Simbel, two hundred miles south of Karnak, origi-
nally faced the rising sun on the date of the thirtieth year jubilee of
Ramses II. The small side chapel faces the winter solstice sunrise.* (Griffith
Observatory, after Gerald S. Hawkins)

REACHING FOR THE SKY

If there is any one image that conjures up all the mystery of Egypt, it
must be the pyramid. The pyramids of Egypt have penetrated deeply
into our consciousness. We find a truncated pyramid on the reverse of
the United States dollar bill and more on packages of Camel cigarettes.
The Great Pyramid at Giza is the one surviving wonder of the original
seven wonders of the ancient world, and it has been a particular attraction for the human imagination. The pyramids have suffered long and
hard at the hands of their interpreters, many of whose ponderous
hypotheses of the true purpose of the monuments seem to outweigh
even the enormous limestone blocks of which the pyramids were built.

The Great Pyramid has been a particular target for pyramid enthusiasts. Among them, Charles Piazzi Smyth, astronomer royal for Scotland
in the mid-nineteenth century, was especially influential. Smyth developed a system of esoteric knowledge based upon what were eventually
proved, by the archaeologist Flinders Petrie, to be erroneous measurements of the Great Pyramid's dimensions. That Smyth was an astrono-

mer and that his "pyramid inch" was brought into disrepute may well have contributed much to the distrust and indifference of archaeologists in subsequent astronomical interpretations of ancient and prehistoric sites.

Whatever controversies may surround astronomical interpretations of the Great Pyramid, at least this much is certain. Its north and south sides are aligned east–west to less than 2½ arc minutes. The west side is aligned north–south to the same accuracy, while the east side deviates most from alignment with cardinal directions: it is 5½ arc minutes west of north. These very accurate alignments imply that the Egyptians were able to make precise astronomical measurements and put them to use in the orientation of at least some of their monumental architecture.

Shafts which lead from the King's Chamber of the Great Pyramid open on its north and south faces. The north shaft is slanted 31 degrees above the level. The Great Pyramid is situated about 1⅓ miles south of the 30-degree parallel of latitude, and this so-called ventilation shaft may have been oriented toward the north pole or, more precisely, toward the upper culmination of the star Thuban, in the constellation Draco.

The other ventilation shaft extends at an angle of 44 degrees 5 minutes from the King's Chamber to the Great Pyramid's south face. It has been calculated by Trimble and Badawy that Alnilam, the central star in Orion's belt, would have transited at this angle at the time the pyramid was built. It is doubtful, however, that either shaft was intended as a sighting tube.

Richard Proctor, a nineteenth-century British popularizer of astronomy, assumed that the Great Pyramid was connected with astronomical observations. In *The Great Pyramid, Observatory, Tomb, and Temple*, Proctor imagined the Great Pyramid put to use as an observatory before it was completed. Truncated at the height of the Grand Gallery's summit, the Great Pyramid, in this guise, would have provided a large, elevated observation platform. The Grand Gallery, Proctor added, was aligned with the local celestial meridian, and its opening to the south could have been used to time the transits of stars. Proctor worked out a clever technique by which precise observations could have been made and an even more clever procedure for astronomically aligning the great structure in the first place.

The Descending Passage extends at an angle of 26 degrees 31 minutes 23 seconds to horizontal, 345 feet from the north face through the pyramid, and into the natural rock on which the pyramid rests. This long passage was designed, according to Proctor, to align with the lower culmination of a polestar. In this way the north–south axis could have

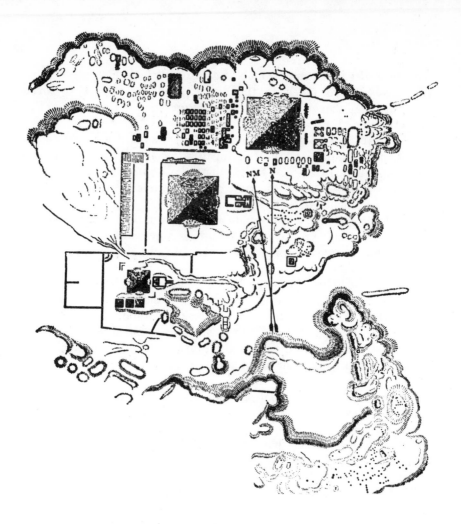

The sides of the pyramids at Giza are accurately oriented to the cardinal directions. (Sir Norman Lockyer)

been determined and maintained. The underground center of the pyramid would be located at the very end of the passage. The approximate center up on the surface could be found trigonometrically. As the structure was built higher, the Descending Passage would eventually become useless for checking the north-south alignment at the surface. Proctor surmised at this point that another passage was prepared to intersect the Descending Passage at the same angle to the horizontal. A small reflecting pool at the junction, observed from the top of the second passage, would permit sighting of the reflected image of the polestar,

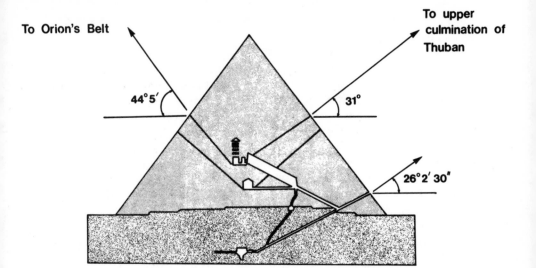

To Orion's Belt

To upper culmination of Thuban

44°5′ 31°

26°2′30″

The Great Pyramid of Giza

The so-called ventilation shafts of the King's Chamber in the Great Pyramid are shown as solid lines which emerge from that chamber at angles of 44 degrees 5 minutes on the south and 31 degrees on the north. The Descending Passage also emerges from the north face, at an angle of 26 degrees 2 minutes 30 seconds. The "ventilation shafts" appear to have astronomical significance and may have been intended to conduct the soul of the dead pharaoh to the circumpolar stars in the north and to the stars of Orion's belt in the south. (Griffith Observatory)

whose light still passed down the Descending Passage. Just such a shaft, the Ascending Passage, intersects the Descending Passage 60 feet from its entrance. The Ascending Passage mirrors nearly exactly the angle of descent of the downward shaft. Thuban would have been close enough to the line of sight of the Descending Passage in 3440 B.C. and in 2160 B.C. to be used as the sighting star. It is estimated, however, that the Great Pyramid was built at about 2600 B.C., and Thuban would not have worked at that time.

Much has been made of the behavior of shadows cast by the Great Pyramid. Proctor's interpretation of the Grand Gallery as a transit also allowed for determination of solstices and equinoxes by shadows cast upon its floor. Moses B. Cotsworth, a British advocate of calendar reform, late in the nineteenth century, fabricated an elaborate theory in which the shadows cast by the Great Pyramid permitted the Egyptians to measure the length of the tropical year to one hundred-thousandth of a day, or approximately to 1 second. There is no textual evidence that Egyptians were equal to this accuracy. Cotsworth also argued that the first disappearance of the shadow from the north face in early spring, when, one day, the sun climbs high enough at noon to illuminate the northern side, could have been used to establish the length of the tropical year. Once again, however, there is no textual evidence to support this notion. Although it seems certain that the Great Pyramid was astronomically oriented, its use as an observatory is unsupported. Its precise alignment must, therefore, have been symbolic of some principle that was important to the Egyptians. Unfortunately, they did not share it with us.

THE RETURN OF THE PHOENIX

The cult of Osiris in Egypt dates back to the First Dynasty and appears to have its origin in the agricultural cycle. Fertility rituals were associated with Osiris, who, like the vegetation, dies until revived in the next season. In Egypt this pattern was closely associated with the behavior of the Nile, and it is not surprising that in some of his early forms Osiris was identified with the Nile.

As Osirianism evolved, the cult became more involved with the cycle of death and rebirth, in which the Egyptians believed. Osiris became the triumphant and transcendant revived god, the ruler of the kingdom of the dead. All Egyptians shared in the hope for immortality in the realm of Osiris. This emphasis is most evident, of course, in the pyramids, which were built to house dead pharaohs and their belongings.

In the Pyramid Texts the dead king was identified with the morning star. Usually this title was applied to Venus. But it may suggest resurrection also, for the decans reappear as morning stars. Their behavior is curiously linked to the preparation of the body for the afterlife.

Sirius and the other decans, which were chosen in the pattern of Sirius, disappear for a period of seventy days before they first appear again at heliacal rising. During this period they were said to be in Duat, the underworld. Otto Neugebauer described a subsequent commentary, in the simpler demotic script, on the hieroglyphic inscriptions in the tomb of Seti I as saying that each decan dies and is purified in the embalming house of Duat, from where, after seventy days of invisibility, it is reborn. We know also that the Egyptian embalmers took seventy days to prepare a body for burial. The language of the stellar cycle appears to be interchangeable with the language of funeral rites.

Sirius and the other decans may also have been linked with another symbol of rebirth, the phoenix. The Egyptians attached particular importance to a heronlike bird they called the "bennu," and the Greeks identified the bennu with the phoenix. According to Herodotus, the red and gold bennu was reported to return from Arabia to Heliopolis after five hundred years absence, or, more curiously, according to others, after 1,461 years. In some versions of the legend the bennu dies at Heliopolis, and from either its nest or its own burned remains a new bennu arises to start the cycle of life anew.

The bennu must have been associated with the sun, for the bird, in Egyptian myth, was said to be the soul of Ra. Later commentators, at least, have regarded the bennu's cycle of death and resurrection as an allegory of the daily revival of the sun on the eastern horizon. The bennu was said to alight, at dawn, upon the "benben," which was a pyramid-shaped object or perhaps an obelisk. This benben was a symbolic representation of the original Benben, the primeval hill, which, in Egyptian cosmology, first emerged from the waters of chaos to become Atum, the God of Creation.

Contradictory aspects of the myth of the bennu and the very idea of the benben have made it difficult to understand the relationship between these two very important symbols and the Egyptian solar religion. Some of these difficulties have been resolved, however, by interpreting the bennu legend as an allegory of the heliacal rising of Sirius. The fire in which the bennu dies is the glow of the sun, which eventually makes it impossible to see Sirius during the period it is a daytime object. Sirius dies when it begins its period of invisibility; it is reborn at heliacal rising. The bennu's return from the Arabian Desert refers perhaps to the reappearance of Sirius in the east, toward Arabia. One of the intervals between returns of the bird to Heliopolis was 1,461 years. This appears to be simply a record of the Sothic cycle and the length of time, in civil calendar years, for the heliacal rising of Sirius to coincide once again with the New Year. The bennu was the guide of the gods in Duat and came from the heart of Osiris. Sirius, or any heliacally rising star, might, therefore, be the revived bird that heralded the sun. Venus, as a morning star, might have taken this role also.

It would not be proper to carry any interpretation too far. The associations between the sun, Sirius, the bennu, Isis, Osiris, and Ra may be too intertwined to extract the original meaning of the myths. We do know that Isis, the wife of Osiris, was strongly identified with Sirius. Osiris and the stars of Orion are also equated. Yet Osiris is associated with the setting sun, too. Lockyer was prompted to generalize that Isis stood for "anything luminous to the eastward heralding sunrise." Osiris became "any celestial body becoming invisible." These broad conclusions have been ignored by most Egyptologists, perhaps rightly, but they do retain an inner logic.

One last peculiarity of the Egyptian treatment of Sirius appears to link it, through the cult of Osiris, to the pyramids. The hieroglyphic inscription for Sirius, ⟁⬳, included an obvious symbol of a star and two other symbols that may be related to the benben. The half circle was used to signify the benben. The long, thin triangle is most reminiscent of a pyramid, or perhaps an obelisk. Margot Wood, a California teacher and writer, has suggested that the benben might once have been an observation platform, from which Sirius and other decans could have been followed. Whether observation of Sirius is the source or the result of the myth of the benben, the link is still clear. Just as the benben symbolized the emergence of existence from non-existence, of the birth of the world, so Sirius, as the bennu, recommemorated creation by alighting upon the benben, be it obelisk, observation platform, or pyramid. It heralded that universal symbol and source of creation, the sun.

The pyramids of Egypt at the time of the Nile's flood are romantically portrayed in this nineteenth-century engraving. (Sir Norman Lockyer)

We need not doubt that the pyramids were tombs for the pharaohs, but their astronomically accurate alignment may be an echo of whatever structures were used to make observations of the heavens. The allegory of Osiris might then be completed for the entombed, who was buried where the "guide of the gods" might alight and resurrect his soul for eternity. The Pyramid Texts describe the ascent of the departed king to the sky. He joins Orion (Osiris), and Sirius is his guide. They continue together as participants in the cosmic cycle. A similar wish is expressed in other texts. The spirits of the dead hope to join the never-setting, never-dying circumpolar stars. These two possible transfigurations, in which the dead pharaoh joins Osiris or the circumpolar stars, may explain the orientation of the so-called air shafts from the King's Chamber in the Great Pyramid. They may be ramps by which the dead king makes his way to heaven.

There may be no need to try to connect the pyramid and the benben with the sun, as has so often been done with unsatisfying effect, for the pyramid may be the agency for rebirth of the king, just as the decans themselves are reborn, as the Pyramid Texts say:

> The King is dead.
> He flies as a bird and settles as the beetle
> Upon the empty throne that is thy boat,
> O Ra.

> The King is dead.
> He has left the earth
> He is in the sky.

> He rushes like the Bennu against the sky.
> He has kissed the sky like a falcon.
> He has leaped skyward like a grasshopper.

CHAPTER SEVEN

Observatories of the Gods and Other Astronomical Fantasies

E. C. KRUPP

Among the sciences, astronomy has been fortunate to enjoy sustained popular support. The objects and places that fall within its authorized domain are, without exception, exotic and distant. Since all of the universe is fair game in astronomy, those who study it immerse themselves in fabulous realms and consider mind-quaking possibilities. The cosmos is populated, after all, by things whose character and behavior exceed even the wildest ravings of the most maddened cranks. We find in the universe both worlds hundreds of times the size of our own and stars not much larger than our major cities. We observe whole galaxies explode with no inkling of the nature of the forces loosed within them.

There is a built-in romance to astronomy that is fortified, oddly enough, by its solid scientific basis. Those who partake of its pleasures find a sense of the cosmic and the unknown that can still be well integrated with the workings of a rational mind.

Archaeology has also had more than its share of "lost continents" and "lost tribes of Israel." And this, too, is not surprising, for archaeologists piece together what for many is the most interesting puzzle of all: how we as humans have come to be the way we are.

Add to the cosmic mystery of astronomy our incomplete knowledge of our origins and the mute remains of monuments like Stonehenge or the lines drawn by the Nazca people on the plains of Peru and you evoke visions of the unknown which are irresistible.

It is, no doubt, the dual appeal of archaeology and astronomy—both their science and their romance—that makes them particularly susceptible to the assaults of cranks and pseudo scientists. After all, if astronomers and archaeologists may luxuriate, sometimes, in fanciful speculation, should they not greet with enthusiasm the animated extrapolations of others? There is a ringer in this, of course, or else the scientific community would have long since embraced those of the fanciful fringe as comrades in spirit if not in arms. At issue in the notions of Immanuel Velikovsky, Erich von Däniken, and others is not their ideas but the way in which their ideas are defended. In matters of belief, the scientific seal of approval is hardly necessary, but when the name of science is invoked in support of "worlds in collision" or "ancient astronauts," the data must be handled in accordance with the rules of scientific evidence. It is on this basis that they should be contrasted with the material that has so far preceded them in this book.

WHEN WORLDS COLLIDE: THE VELIKOVSKY CATASTROPHE

In 1950 Immanuel Velikovsky, a psychiatrist, published his book *Worlds in Collision* and created a stir in the astronomical community that has really yet to subside to the level the book deserves. Through an ill-advised attempt at indirect censorship, the astronomical establishment succeeded in generating high public interest in a book which, because of its plodding style and abstruse content, might otherwise have slipped into obscurity. The astronomers became identified as the villains of the piece, for it appeared as though the true spirit of scientific inquiry was stifled by a new Inquisition. To some extent this occurred, but Velikovsky then wrapped himself in the cloak of Galileo and attracted supporters eager to leap to the defense of his beleaguered ideas. They have been substituting self-righteous rage for rational argument ever since. Their hostility is no better, and probably no worse, than that of the astronomers who condemned the book without reading it.

The theme is set early on, in the preface to *Worlds in Collision.* There Velikovsky says, in reference to Newton and Darwin, "If these two men of science are sacrosanct, this book is a heresy." Velikovsky has chosen to sidestep known evolutionary processes in nature in an attempt to explain conditions in our solar system in terms of catastrophes that have taken place within the historical record of human events. Although his subject is the planets, his arguments rely upon myths and legends of many ages and peoples. Into this catalogue of clues he cunningly inserts the lead player in his drama—the comet.

Velikovsky reminds us of the zone of minor planets between the orbits of Mars and Jupiter, and we are asked to recall that this debris might once have formed a planet long since destroyed. This morbid notion still finds its way into serious discussions of a genuine mystery, the origin of the minor planets. Although most astronomical evidence does not favor the idea that a once-whole planet exploded into tens of thousands of fragments, which continue to orbit the sun, the theory has not been completely ruled out. Velikovsky is not really interested in the subtle details of these arguments, however. He is content, simply, to create an atmosphere imbued with cosmic catastrophe. He asks what might have destroyed that hapless planet and suggests perhaps a comet.

The reputation of the comet as a harbinger of disaster is long and honored. Comets have been regarded as portents of war, famine, and destruction throughout history. A medieval prayer asked God for protection "from the comet and the fury of the Norsemen." Velikovsky provides us, then, with the traditional messenger of divine wrath having already destroyed a highly unlikely planet between Jupiter and Mars. It is a short leap of faith from this point to the transformation of the divine wrath of the God of the Hebrews into a comet whose side effects were nothing less than the plagues on Egypt, the parting of the Red Sea, and the halting of the sun and the moon over the battle of Gibeon.

Velikovsky was not the first to exploit a comet in the name of terrestrial calamity. Ignatius Donnelly, a mid-nineteenth-century United States congressman from Minnesota and erratic chronicler of the former glory of the sunken continent of Atlantis, had proposed that comets were responsible for a variety of past catastrophes. Before him, the English theologian and mathematician William Whiston, at the end of the seventeenth century, related the biblical flood to a comet's close passage by the earth.

According to Velikovsky, a comet was belched forth from the planet Jupiter. After several near encounters with the earth, at about 1500 B.C., the comet settled into a nearly perfectly circular orbit around the sun, lost its tail, and became the planet we know today as Venus. A comet in near collision with the earth would, Velikovsky tells us, distort the earth's motion and cause a rain of meteorites upon the surface of the earth. Doubtless the planet Venus would do the former, but neither Venus nor a comet is likely to storm our planet with meteorites.

Perhaps we ought to examine the nature of comets. Mostly, they are bluff and fancy. A comet is a snowball, ices of gases frozen at the outer edges of the solar system. It may therefore be the one thing in the cosmos that is consistent with Hanns Hörbiger's cosmic eternal ice theory. Hörbiger was an Austrian engineer who, in the early decades of this century, developed a pseudoscientific explanation of Atlantis and cosmic catastrophe. More about this will be mentioned later.

A comet is a small object, by no means planet-sized. The actual solid nucleus of a comet is at most one millionth the mass of the earth (and probably a hundred thousand times less than that). The diameter is just a few miles across. When a comet returns from the far end of its elongated orbit to the inner solar system, it is heated by the sun. Volatiles in it evaporate to produce a huge head, or *coma,* which can be many times larger than the earth, and a tail millions of miles long. But this vast size is deceptive; it's all show. You could fit comfortably all of the atoms in a comet's tail into a suitcase, and it's an old astronomical adage that a comet is the nearest anything can be to nothing and still be something. Its mass is so low that a comet, passing by the earth, would have its motion distorted, but the comet itself could hardly affect the earth.

Velikovsky's Comet Venus was really something special. In his scenario, after explosive ejection from Jupiter, Venus nearly collided with the earth. The first encounter took place at the time of the Exodus, and Venus appeared as smoke by day and fire by night. It dropped red dust, frogs, lice, flies, locusts, vermin, hail, burning oil, manna, and meteorites on the earth. It caused the eruption of boils, the plague, darkness, and earthquake. A second encounter took place fifty-two years later, and after that Comet Venus settled down into the nearly perfectly circular orbit it follows today. In permanent orbit the maverick's tail was bobbed.

Astronomers wince at the thought of red-hot meteorites raining down on earth from the incandescent surface of a Comet Venus. It is a popular misconception that meteorites are hot. Even today observations of a bright meteorite are often accompanied by reports of fires it was thought to have started. The usual terminology for an especially bright meteor, "fireball," doesn't help matters much.

A piece of interplanetary debris is known as a "meteoroid," and as it passes through the atmosphere, it heats the air around it by compression, and this glowing air is the "meteor" that is seen in the sky. The outer surface of the meteoroid is heated also, but internally it remains as cold as interplanetary space. Most of the object is ablated away, but any portion that survives the fall to strike the ground is termed a "meteorite," and within the first thirty minutes or so after its fall it may feel warm to the touch. Meteorites have even been described, however, as forming a crust of ice, for by the time the heat of passage has es-

A real comet, like Comet West, which appeared and was photographed in the spring of 1976, bears little resemblance to what Velikovsky's Comet Venus was like, had it existed. (Ronald E. Royer and Steve Padilla)

caped, the object's surface is bitter cold, like its interior, and water vapor can condense on it and freeze.

Furthermore, meteorite falls do not coincide with the passage of comets. Meteorites and comets follow different orbits around the sun and, in general, are unrelated.

Velikovsky also tells us that comets are made of carbon and hydrogen. They can contain no oxygen, however, he says, for if they did, they would ignite and burn in space. According to Velikovsky, if the tail of a comet passes through the earth's atmosphere, the tail will light into sheets of flame. In 1910 the earth passed through the tail of Halley's

Comet. No meteorites rained to the earth in flames. No fire was ignited in the sky.

The principal luminous constituents of comets are know to be carbon monoxide, carbon dioxide, and nitrogen. It would be difficult to imagine a less flammable mixture. It is also a far cry from the petroleum, or hydrocarbons, Velikovsky insists fell from the comet in torrents on earth. From these hydrocarbons he derives, in a transformation more miraculous than water to wine, his carbohydrates. The original transformation of hydrocarbons to carbohydrates in *Worlds in Collision* occurs so subtly it goes nearly unnoticed. There is no explanation to accompany the shift in words, no argument to make the point clear. Velikovsky needs the carbohydrates to rain upon the earth (except on the seventh day) as manna for the Israelites. Velikovsky's ignorance of chemistry apparently permitted him to make this shift from fuel oil to ambrosia without a blink. In recent years he has called upon bacterial action to effect the change, although he apparently did not understand the chemistry of the two substances when he wrote *Worlds in Collision*.

In any case, the rest of a comet's appearance is governed by the reflection of sunlight off the dust that escapes from it. Analysis of infrared light reflected from the clouds of Venus has discounted the presence of either hydrocarbons or carbohydrates in the atmosphere of that planet. Velikovsky, in turn, rejects the best current results and has called for more definitive evidence.

A comet, by the way, is not hot. The atoms which evaporate from its nucleus may be at high temperature, but there are so few of them, there is no heat to speak of. Yet this alleged cometary heat is the source of Velikovsky's much-touted "prediction" of the high temperature of Venus. Despite the fact that comets are not hot, Velikovsky associates comets with red-hot stones and sheets of flame and very loosely suggests that at least some of them may have been ejected from planets in relatively recent times.

Velikovsky is impatient with the astronomers who hold him to the facts about comets. Comet Venus, after all, is only compared with a comet because of its alleged appearance in the sky as it was passing close to earth. It is fair, however, to take Velikovsky to task for what he might regard as a matter of semantics, for he invokes the emotionally loaded reputation of the comet and also uses what known scientific facts about comets he needs to punctuate his message.

Venus is imagined by Velikovsky to be an ejected comet which, through adolescent collisions with the adults of the solar system, was heated to incandescence. To be sure, Venus presents us with problems, but then so do most objects in the universe. Their mysteries and contradictions are often what motivate scientific inquiry. At the time *Worlds in Collision* was being written, it was known that the side of Venus that faced away from the sun was at about the same temperature as the daytime side. Venus was also known to rotate slowly, and it was difficult to reconcile comparable nightside and dayside temperatures with a planet which, by turning so slowly, ought to cool off on the nighttime side. Velikovsky explained this apparent contradiction by predicting that Venus itself is hot, incandescent perhaps, and the release of this heat, and not the energy from the sun, keeps all sides of Venus uniformly hot. We now know that the upper atmosphere of Venus circulates very rapidly, and the initial temperature measurements obtained by astronomers referred to the upper clouds.

Venus did turn out to be surprisingly hot at the surface, nearly 900 degrees F. This high temperature may be at least partially explained by the runaway "greenhouse effect." In this process, solar energy penetrates through the atmosphere and is absorbed by the planet. Much of this energy is reradiated as infrared energy, or heat, but this radiation cannot pass back through the carbon dioxide atmosphere. The greenhouse effect works on earth, too, but the temperatures are cool enough here to permit liquid water to exist. The earth's water helped convert much of the original carbon dioxide to carbonates, but on Venus the water remained gaseous. The water vapor and the carbon dioxide were compelled to rise to higher and higher temperatures in a runaway effect.

Velikovsky objects that the runaway greenhouse theory in its present form is not completely adequate. Microwave observations indicate there may not be enough water vapor in the atmosphere of Venus to do the job. But even the abandonment of the greenhouse explanation would go no further to endorse Velikovsky's explanation. His theory, like any other, must stand upon logical hypotheses, must be supported by relevant arguments, must be consistent with known facts, and must make predictions which can be tested.

It is difficult to evaluate Velikovsky's hypotheses through consideration of the myths and legends upon which his ideas are based. The stories are plucked from their original contexts and languages. The sheer mass of references, presented with all of the traditional trappings of scholarship, cannot help but strongly affect the unbiased yet uncritical reader. Fortunately, Velikovsky does reach some specific conclusions, and a careful review of their possible validity can help tell whether his guesses, when right, are the result of chance or genuine science.

Velikovsky tells us that Comet Venus profoundly affected the earth. Mars, he says, was also disturbed by the erratic path of Venus in the

The upper atmosphere of Venus was revealed by Mariner X to circulate with an elegant spiral pattern. (NASA, Jet Propulsion Laboratory)

eighth century B.C., and as a result, Mars also passed dangerously close to the earth in 721 B.C. A second near miss in 687 B.C. was responsible, he adds, for the destruction of Sennacherib's army. These near-collisions of Venus and Mars with the earth

a. reversed the earth's magnetic field;
b. changed the rate of the earth's rotation;
c. changed the geographical location of the poles;
d. changed the period of the earth's revolution around the sun; and
e. changed the alignment of the earth's axis in space.

Apart from any consideration of any logic present or absent in Velikovsky's drama, it is possible to look for evidence of the alleged changes in the earth's orientation and motion. Contradictory evidence could permit us to discard Velikovsky's script.

Paleomagnetism of rocks at the bottom of the Atlantic show evidence of reversals of the earth's magnetic field. These shifts occurred periodically and can be dated. The last shift took place about 30,000 years ago, and this is far too early for the reversal which Velikovsky claims happened in 1500 B.C.

Many arguments against a slowing or stopping of the earth's rotation rate have been voiced. Perhaps the most effective of these is that of the American scientist-author Isaac Asimov. He rightly notes that Velikovsky's rapid change in the earth's rotation rate, which allegedly took place no more recently than the seventh century B.C., should have snapped the ends off the long, tapering, and brittle stalactites and stalagmites in the Carlsbad Caverns. These delicate structures require tens, even hundreds, of thousands of years to form, and they are still intact.

Almost any shift of the earth's crust relative to the geographical poles will change the directions of north, south, east, and west. Only the highly unlikely slippage of 180 degrees, in which the north and south geographical poles simply exchange positions, will keep a structure oriented on the cardinal directions. The pyramids of Giza undeniably were built before 1500 B.C. and Velikovsky's first encounter between earth and Venus. They now are aligned accurately with the cardinal directions. If we accept Velikovsky, we must conclude that the pyramids were originally so aligned and that the shifting of the geographic poles has been exactly 180 degrees or we must conclude that the pyramids were randomly oriented and brought fortuitously into alignment. Neither option seems probable. Professor Alexander Thom's many equinoctial alignments of Megalithic sites also argue against a shift of the geographic poles. Velikovsky has subsequently modified his view about geographic shifts and now claims that although they occurred, they were only temporary.

A shift of the celestial poles would simply point the axis of the earth in a different direction in space. A change of this sort is also hard to support. Archaeologist E. W. MacKie has noted that Thom's precision solar observatories—for example, those at Ballochroy and Kintraw, pin down the obliquity of the ecliptic for an epoch prior to 1500 B.C. The stones were set up no later than 1800 B.C., and they accurately indicate, through distant foresights, the setting points of the sun at winter and summer solstice for that epoch. Only a 180-degree change in the position of the celestial poles could duplicate the precise solar alignments. Any other change would require different foresights for the geographical locations of the various solar observatories.

The Great Pyramid at Giza provides another confirmation that the orientation of the earth is much as it was before 1500 B.C. After known corrections for precession are made, the so-called air shafts of the Great Pyramid align with the circumpolar star Thuban, in Draco, and the stars of Orion's belt. Textual evidence supports the significance of these directed shafts. A shift in the geographic pole would destroy the northern shaft's alignment. A shift in the celestial pole would ruin the south's.

Velikovsky has used the astronomical ceiling of the Tomb of Senmut to justify his claim that the celestial poles have reversed and that the direction of the earth's rotation has reversed. The southern panel of the ceiling depicts Sirius and the stars of Orion so that they are upside down when the viewer faces south. According to Velikovsky, this seemingly awkward arrangement records a time when the poles and direction of earth's rotation were reversed from what they are now. This supposedly resulted in the present arrangement in which the stars of Orion appear to be east of Sirius and to precede it. There is no reason to suppose, however, that the stars on the southern panel were intended to represent the sky as it would be seen overhead of the viewer from inside the tomb. Indeed, if this were intended, even as Velikovsky suggests, we would expect the figures of Osiris (Orion) and Isis (Sirius) to be right side up, for why should they be imagined as upside down from earliest times until the cataclysm? Even more significant, the hieroglyphics are also upside down if we assume Velikovsky knows how one is to stand and observe the ceiling. By contrast, if we do not require the ceiling itself to be oriented astronomically and if we define what direction is intended to be up by the inscriptions, the constellation figures are perfectly in accord with their appearance in the sky.

The number of days it takes the earth to orbit the sun can be changed in two ways. If the length of the day is altered, the number of days in the year must change. Additionally, a change in the actual time, measured in hours, say, to complete an orbital cycle must be accompanied by a change in the distance between earth and the sun. Velikovsky is vague on the exact details of the changes he claims occurred in the length of the year, but he suggests both types of change took place.

Velikovsky implies that the earth was closer to the sun prior to 1500 B.C. He does not appear to be any more specific than this, but some of his supporters have concluded that the semimajor axis of the earth's orbit was about 81 per cent of what it is now. Such a size would result in a year of 317 twenty-four-hour days.

MacKie offers proof that the earth at that date was not closer to the sun than it is now. Thom's alignments at Ballochroy and Kintraw are in agreement with the setting of the upper limb of the sun at summer and winter solstice respectively. Both alignments, if corrected for the apparent radius of the sun, give the same declination for the sun's center. This is an impressive confirmation of the astronomically precise character of these two independent sites. It also permits us to deduce

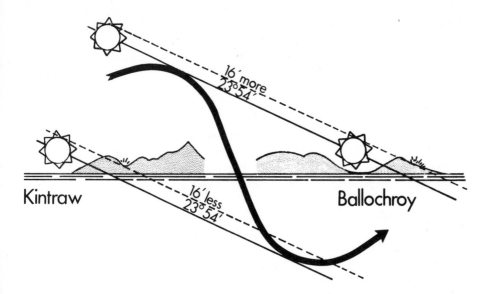

We can see in the upper dashed line the path of the upper limb of the setting sun at the winter solstice as seen from the Ballochroy stones on
Kintyre. A second dashed line indicates the path of the upper limb of
the setting sun at the winter solstice as seen at Kintraw, in Argyll. Both
solar observatories provide extremely precise alignments on the sun and permit us to estimate the angular radius of the sun as it was seen before 1500
B.C., Velikovsky's date for his catastrophe. The sites are older than 1500
B.C., and we know that over the course of a year the center of the sun oscillates in declination between the two extremes marked by solid line. The
wavy arrow indicates this annual movement in the declination of the sun. To
obtain the declination of the sun's center at Ballochroy we must subtract 16
arc minutes. At Kintraw, to obtain again the correct declination of the sun's
center, we must add 16 arc minutes. These two observatories require addition on the one hand and subtraction on the other of the same number to
obtain the correct extremes of the sun's movement for the time at which
the observatories were in operation. This fact not only supports the hypothesis that these sites were really observatories, but it also pins down the radius and hence the angular size of the sun as seen from earth. It was the
same then as it is now, in contradiction with Velikovsky's cataclysmic
changes of the distance from the earth to the sun. A related argument
shows that the polar axis was oriented in pre-Velikovskian cataclysm times
as it is now. (Griffith Observatory, after an original drawing by Euan W.
MacKie in *Pensée* magazine, Portland, Oregon)

that the apparent diameter of the sun today, 32 arc minutes, was the apparent diameter before Velikovskian calamity. If the sun appeared to be the same size, it was the same distance from the earth. The orbital period could not have changed.

Many more arguments could, and have been, marshaled against *Worlds in Collision*. The moon, for example, must have been badly jostled in its orbit by the alleged near-encounters of Venus and Mars. Archaeoastronomy provides us with sites like Temple Wood, in Argyll, however, which antedate the Venus collisions and yet which are accurately aligned on significant moonsets, as we might expect had no collision occurred at all. MacKie did offer that the Megalith Builders may have responded to some natural catastrophe that predated the construction of the monuments around 1800 B.C. Their intense interest in a revised celestial order might "neatly explain" the period of intensive building. Of course, the Megalith Builders would have needed similar Megalithic instruments prior to the catastrophes, or they would have had little observational basis for noting the change. MacKie may have been trying to soothe any controversy his other remarks may have stirred, for his conclusions were published in the pro-Velikovsky journal *Pensée*.

Even Stonehenge I, in Wiltshire, with its recalibrated radiocarbon date of at least 2800 B.C., includes astronomical alignments that are understandable only in the absence of Velikovskian catastrophes.

Ancient and prehistoric dates are another source of trouble for Velikovsky, for much of his "evidence"—the celebrated papyrus of Ipuwer, for instance, and much of the historical framework of his libretto for disaster, such as the Exodus—are assigned dates in considerable conflict with those accepted by scholars. Velikovsky admitted that he uses "a synchronical scale of Egyptian and Hebrew histories which is not orthodox." One whole book, *Ages in Chaos*, Volume 1, was devoted by Velikovsky to reconstruction of history that would square with his sequence of cataclysms in historical times. Velikovsky compressed Egyptian history to suit his needs, but by the end of the book he had still failed to reach his own goal: the age of Alexander. Instead, Velikovsky has a thousand years to go and six hundred "phantom" years of Egyptian history to eliminate during that period. The final solution was promised in 1952, but what was then the soon-to-be-published *Peoples of the Sea* (the second volume of *Ages in Chaos*) did not see the light of day until early 1977. It will remain for historians to evaluate Velikovsky's answer to his chronological dilemma.

Persistence against such impossible odds seems hopeless, but Velikovsky clothed himself in the vestments of the heretic, a posture which through fervor can give strength. If, as he stated, Newton and Darwin are sacrosanct, then he is the renegade. This is certainly so, for early in *Worlds in Collision* Velikovsky describes the process whereby planets orbit the sun. There are, he tells us, two forces. The sun's pull on the planet is one. The other is the planet's centrifugal force away from the sun. Were it not for this second force, Velikovsky emphasizes, the planet would fall into the sun.

Well, there is no second force, and the planet, as always, is falling around the sun. The path of this fall is the planet's orbit. Velikovsky not only rejects Newton, he fails to understand him.

In the face of unabated criticism, Velikovsky and his partisans called for atonement. The shame of his treatment at the hands of the astronomical establishment, in their eyes, could only be redressed by a fair hearing before the scientific community. The essential elements for such an event were mobilized on February 25, 1974, in the form of a special symposium of the American Association for the Advancement of Science, dubbed "Velikovsky's Challenge to Science." Both sides undoubtedly hoped that the public debate would, once and for all, end the controversy. Velikovsky's critics expected to relegate *Worlds in Collision* to the files of crank literature forever. Perhaps his partisans hoped to see the opening of an era of serious study of Velikovsky's ideas and, at the very least, vindication where vindication was due. The critics were devastating; Velikovsky remained unbowed.

LEY OR NAY?

Prehistoric monuments and ancient curiosities cover the countryside in Britain. The Ordnance Survey maps record, in terms of the national grid reference system, the locations of thousands of monuments, and to a person tramping with map in hand through the Wessex Downs or across the open moors of Devon it is quickly demonstrated how long-inhabited is the British landscape and how domesticated it has become. Once the eye is adjusted to the subtlety of the surrounding countryside, Bronze Age barrows, Iron Age hill forts, Neolithic standing stones, dikes, Roman roads, and other antiquities seem to dominate the countryside, like a secret government.

The search for and visit to an ancient site, or a view of it first seen in the distance, always has an emotional impact. Ruins seem saturated with mystery. Who built them and why? The great antiquity of archaeological monuments and their enigmatic appearance conjure images of people long gone and motives long forgotten.

Since 1724, when William Stukeley published his account of his travels in Britain, *Itinerarium Curiosem*, and probably before, antiquarian pursuits in Great Britain have been fashionable. Many, like Stukeley, have not been content simply to observe and record what they have seen but have indulged in wild speculation. Stukeley, without solid archaeological evidence, argued strongly that the ancient Druids were responsible for many of the old stone circles and that these monuments, in many cases, were sites at which an old serpent worship was practiced. Since Stukeley's time, prehistoric Britain, like the pyramids of Egypt, has been repeatedly reinterpreted in terms of one strange idea after another. One of these notions, "the old straight track," is enjoying a revival, to some extent stimulated by Alexander Thom's imaginative and careful analyses of Megalithic monuments.

An "old straight track" is simply a straight line that theoretically connects several ancient or prehistoric sites. On June 30, 1921, Alfred Watkins, a successful English businessman, sensed suddenly, as in a visionary experience, the landscape of his native Herefordshire crisscrossed by a network of straight tracks, or "leys," as he also called them. The lines led from one ancient site to another and often intersected at churches and old stones. Enthusiastic ley hunters have been extracting more leys from the landscape ever since, and nets more complex than the work of the most drug-crazed spider are presented, as a sort of monumental macramé, as evidence of the skill and lost knowledge of the prehistoric people of Britain.

Invisible straight lines across the British countryside need not be dismissed out of hand. There are systems of invisible lines across England today—the National Grid, for example, and the Prime Meridian itself through Greenwich—that are no less real to their users than a system of ley lines could have been to the ancients. The basic question here is whether the ley lines *do* comprise a system and were drawn intentionally between antiquities.

Alfred Watkins suggested that the leys were indeed products of conscious, plotting minds. In his books, *Early British Trackways* and *The Old Straight Track*, he demonstrated the existence of lines marked by mounds, stream fords, stones, wells, moats, and so on. Watkins was convinced that prehistoric people had set out these markers to indicate straight tracks. Often these tracks were sighted on natural horizon features: notches, peaks, and trees. Long-buried trackways along some leys have been excavated. In some cases, churches and crossroads now sit where older, sanctified monuments were once located.

Actually, Sir Norman Lockyer had provided one example of a straight track twenty years before Watkins' books were published. Lockyer had noted that a straight line passed through the centers of Stonehenge, Old Sarum hill fort, and Salisbury Cathedral and lay tangent to Clearbury Ring hill fort.

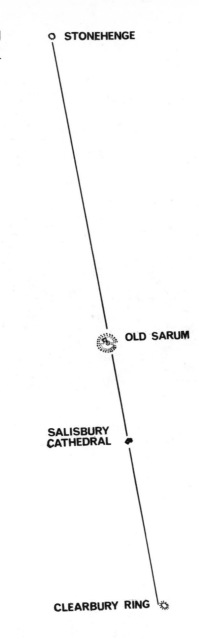

A *classic ley line extends from Stonehenge in the north, through Old Sarum and Salisbury Cathedral, and ends tangent to Clearbury Ring in the south.* (Robin Rector Krupp)

South across the ramparts of Old Sarum the ley line to Salisbury Cathedral reaches the edge of Clearbury Ring, which is in the distance on the horizon and just left of the cathedral spire. (E. C. Krupp)

A ley, then, is just a straight, sighted track. Watkins adopted the word "ley," which in Saxon times meant "a glade or clearing," because it was so often associated with place names of sites situated along the track. Watkins thought that the leys may have been set out as trade routes, along which important commodities, like salt, pottery, and flints, were transported in prehistoric times. Perhaps some of the tracks were actually walked. Others may have been symbolic.

Although Watkins did not investigate astronomical alignments of leys to any great extent, he did incorporate many of Lockyer's solar alignments at Stonehenge and other Megalithic sites into the ley system. The sunrise alignments, according to Watkins, are identical with some of the long-distance leys.

Alexander Thom has argued convincingly that the Megalith Builders were practical-minded and capable of carefully orienting stone rings whose dimensions and shapes satisfied certain prehistoric rules of construction. In addition, he has found evidence not only for symbolic astronomical alignments in Megalithic monuments but also for highly precise solar and lunar observatories. High-precision observatories like Ballochroy and Temple Wood have two major elements: a place to stand and a place to look. The place to stand is marked by a stone, mound, or other Megalithic feature. The place to look may be an artificial marker, but most often it is a distant notch or peak on the horizon. The great distance between foresight and backsight is what gives the observatory its precision.

Thom's interpretations of Megalithic monuments require us to reconsider the Neolithic and Bronze Age cultures that built them and force us to conclude that these early inhabitants of Western Europe were far more than "savages howling in the wilderness," as a television "special" characterized them some years back. This realization and the superficial similarity between a sighted track and a long distance astronomical alignment have revitalized partisans of the ley system. Other unorthodox interpreters like John Michell and Paul Screeton have carried the discussion far beyond Alfred Watkins' original conceptions.

John Michell regards the ley system as just one component of a geomantic tradition as old as prehistory. (Geomancy is the magical art of siting and designing tombs, temples, and other structures according to principles of "sacred" geometry and geography.) In his book *The View over Atlantis* Michell equates the ley system with the ancient Chinese system of divination called *fêng-shui*. The Chinese geomancer used the principles of *fêng-shui* to guide any planned modifications of the landscape into harmony with invisible currents of force (*lung-mei*, or dragon paths) which gave the countryside its character. A sort of geomantic environmental impact report would have been prepared to insure the integrity of the countryside, although here, the principles of design transcended what we might call "mere aesthetic considerations." A unified, integrated design might take into account a variety of concerns: aesthetic, astronomical, mathematical, and utilitarian.

It appears that the Chinese geomancers used natural magnets in their divinations. Site orientation, as we have already seen in the previous chapters, was important to many ancient prehistoric societies. The earth's natural magnetism, roughly oriented as it is with the celestially oriented cardinal directions, may have been regarded as a vital fluid, or life force, with whose currents the human landscape must conform.

Ley lines, in Michell's view, mark the circuits of a subtle telluric energy (that is, earth energy), that is not restricted to, but may be related to, terrestrial magnetism. This telluric energy is an ineffable sort of stuff that physical science has been unable to detect or isolate, but that in no way diminishes the importance of the subtle power to those who believe in its existence. Extravagant claims are made by Michell, in which prehistoric passage graves, like New Grange, in County Meath, Ireland, become energy accumulators. The old straight tracks are alleged to follow water lines, or underground streams, and related lines of force detected by the dowser and his divining rod. Ley lines are regarded as currents of power, the secret of whose use has been lost. Michell in all seriousness suggests that the ancients somehow utilized the power of the "dragon paths" to fly through the air. Bladud the Druid, a legendary figure of the first century A.D., allegedly crashed during a flight on a levitated stone. Paul Screeton devotes a considerable portion of *Quicksilver Heritage* to the connections between spirituality and leys, and according to him, ley lines have considerable esoteric meaning. UFOs are alleged by some to fly along straight lines (see Aimé Michel, *Flying Saucers and the Straight-Line Mystery* [1958]), and these lines were subsequently identified with leys. The occult character of these hypotheses makes them impossible to verify, particularly when the existence of the leys themselves is in doubt.

Watkins' *The Old Straight Track* is not scientific proof of the ley hypothesis. Probably for that reason and because, in the 1920s when the book was published, archaeologists had considerably less respect for the achievements of the Neolithic peoples of Western Europe than they are now developing through new lines of research, Watkins' work was first criticized and then ignored. *The Old Straight Track* is actually quite a reasonable book. But it does not prove the existence of leys. It does begin to outline what might constitute a ley. Despite the enthusiasm of amateurs who have filled the void left by professionals, the ley hypothesis remains unconfirmed. The "earth spirit" literature of Michell, Screeton, and others promises what it cannot deliver, namely, proof that ley lines were intentionally marked on the prehistoric landscape.

Alignments are easy to find. Proof of their significance is difficult. Scientists are trained to handle data scrupulously and according to the rules of scientific evidence. Physical scientists do this well because the phenomena and the data that interest them are the most amenable to this rigorous treatment.

When a scientist is skeptical of a particular hypothesis, the community of interested observers should not fault him or her for being reactionary or obstructive. Part of the job of science is the rigorous, unbiased testing of hypotheses. Because we unconsciously venerate science as the source of authority, we are frustrated when our romantic imaginations come into conflict with that authority. Science, however, is just a way of handling information. It allows certain conclusions with a certain kind of certitude. We may not need the scientist's approval, but if we are going to seek it, we will have to play according to science's rules.

The problem with the ley hypothesis is that it is not defined clearly. It is interesting, in *The View over Atlantis*, for example, to be bombarded with a vast assortment of artifacts and lore, but it is dissatisfying to see them all receive equal weight. All may be evocative and mysterious, but the author provides no means of establishing an over-all, self-consistent vision.

Michell's *The Old Stones of Land's End* is a lovely collection of photographs, maps, and hypotheses, but it does not analyze the ley lines of Land's End critically (although a paper by ley enthusiasts P. Gadsby and C. H. Squire in a recent issue of *The Ley Hunter* argues well that Michell's sites form a close to complete set and that most of the leys identified by Michell are confirmed). Ley lines are imposed by Michell upon the landscape artificially. To play fair, one must define a ley line. What features are to be recognized as markers and why? What is the minimum number of markers to a line? How many lines could be drawn that aren't drawn? How do we know which lines are significant and which are not? What level of probability will we accept of chance alignment? Finally, what is the model of the use or meaning of leys? A sound hypothesis here could suggest tests, which could, in turn, be performed to rule in or rule out a theory. Needless to say, the ley hunters have not fully addressed themselves to these questions, nor have they subjected themselves to this discipline. When they do so, they will get the attention of the scientist.

One British skeptic, Robert Forrest, in a 1975 article, "Leys, UFOs, and Chance," demonstrated how vulnerable the ley hunter is to the effects of chance. In one experiment he showed that on a typical 1:50,000 Ordnance Survey map (approximately 1¼ inches to the mile and depicting an area approximately twenty-five miles square) one can expect to find about 330 antiquities suitable as ley markers. These will generate approximately 1,100 ley lines with at least four markers each. Rarely do the ley hunters find or include these embarrassing riches of evidence.

Alexander Thom's Megalithic astronomical observatories, aligned on distant foresights, may seem to bring us perilously close to acceptance of the existence of ley lines, but the similarity between ley and astronomical sighting lines is misleading. The two ideas may parallel each other, but neither proves the existence of the other. A ley line that turns out to have astronomical significance does not confirm either the ley-line hypothesis or the astronomical hypothesis. Either may be fortuitous. We might ask how such an apparently significant arrangement can emerge from coincidence, but random events play a major role in the behavior of the universe. There is a strong urge to seek and find pattern in observations. The scientific method provides a discipline for reducing the effect of subjective judgment to an acceptable limit.

What makes Thom's alignments interesting is not merely the fact that they are astronomical but that his hypothesis is fairly narrow. Alignments are defined to a specified (and high) precision. This rules out many sites that a less choosy investigator might believe real. The Thom sites share certain elements in their construction and design. There are enough carefully surveyed sites to make a meaningful comparison. Finally, alignments are not merely postulated. Rather, a practical, detailed method for using the sites is described. This evidence is necessarily circumstantial, but it is particularly compelling at the lunar sites, where remnants of what may have been very necessary pieces of auxiliary equipment, the extrapolation triangles and grids, are found.

There is much more to the earth spirit canon than can be detailed here. According to Michell's *City of Revelation*, venerable monuments like Stonehenge and Glastonbury Abbey are built to the same principles of sacred geometry and what is called "gematria." The ancient art of gematria simply assigns a number to each letter of the alphabet; any word or phrase therefore assumes a numerical value, and these values carry an esoteric, symbolic message. The gematria of Stonehenge, for example, becomes a contrived combination of dimensions and geometric relationships in which certain significant numbers are contained. From a magic square traditionally identified with the sun, several important numbers emerge, for example, 74, 370, 666, and others. Michell reports that the circumference of the bank at Stonehenge is 370 Megalithic yards. The Aubrey Holes encircle an area of 6,660 square English yards. A hexagon inscribed within the bank has an area of 7,400 square English yards. Two intersecting circles, each of diameter 666 English feet, drawn on the diameter of the bank, produce by their intersection a *vesica*, whose maximum width equals the diameter of the Sarsen Circle, which is centered within the *vesica*.

Michell has allowed himself considerable latitude by invoking whatever geometric dimensions and units as will coincide with the mystic numbers. This analysis of sacred geometry is completely subjective and omits many possibilities which another investigator might find significant. Here, as with ley lines, we find it very difficult to test the hypothesis. This does not mean that prehistoric peoples did not set out the old straight tracks. Indeed, they had the ability to do so. Whether they actually did set them out is another matter, and it awaits proof.

THE GLASTONBURY ZODIAC

Glastonbury Abbey, like Stonehenge, gets a complete interpretation in terms of gematria and sacred geometry. This is not especially surprising, for Glastonbury has remained an important spiritual site throughout the centuries. Pilgrimages to it continue to be made, and many legends surround it. Glastonbury, in Somerset, is traditionally the location of the first Christian church in Britain, allegedly built in the first century A.D. by Joseph of Arimathea. Arthurian legends, including the quest for the Holy Grail, are part of the lore of Glastonbury, and Arthur and Guinevere are said to be buried on the grounds. An even more amazing claim was made by an Englishwoman, Katherine E. Maltwood, in 1929. She discovered what she claimed was a set of huge representations of the figures of the zodiac, spread out ten miles across the Somerset landscape.

Mrs. Maltwood detected the enormous figures on aerial survey photographs. To her eye, the natural features of the land, hills, rivers, and such, and artificial constructions in the form of ditches, banks, roads, and field boundaries outlined twelve figures, arranged more or less in a circle, and a thirteenth, the so-called Great Hound of Langport. Actually, Dr. John Dee had referred about three and a half centuries earlier to a pattern of earthworks in the vicinity of Glastonbury that was laid out in the form of a zodiac. Mrs. Maltwood was unaware of Dee's remarks when she published, first anonymously, her *Guide to Glastonbury's Temple of the Stars,* and it is still not known whether Dee was referring to monuments on the same scale as Mrs. Maltwood's.

Mrs. Maltwood's "Giants of Somerset" may seem to share a kinship with the Amerindian Adena culture's effigy mounds (see Chapter 4) and with the huge figures drawn by the Nazca peoples on the plains of Peru, to be discussed later, but the Glastonbury Zodiac figures are much larger. Mrs. Maltwood claimed that superposition of a map of the figures upon a celestial planisphere, drawn to the right scale, shows a good agreement between the relative positions of the earthworks and the positions of the stars on the sky map.

Although some of the Somerset landscape figures do overlap the appropriate star groups in the combination diagram Mrs. Maltwood provided with her book, it would be difficult to avoid some matches. In any case, actual stars do not coincide with any special features on the ground—the star markers, so to speak. The relative sizes of the terrestrial figures do not match the celestial zodiac either. For example, the Glastonbury ram is much too large for Aries.

All of the Glastonbury figures are said to have their heads turned toward the west, and Mrs. Maltwood regarded this arrangement as fur-

The figures of the giant Glastonbury Zodiac do not exactly correspond to the actual sidereal zodiac. (Robin Rector Krupp, after K. E. Maltwood)

ther evidence that the Glastonbury Zodiac was real. She was quite interested in associating incidents in the Arthurian quest for the Holy Grail to the zodiac figures, and many of her arguments consist of mixed collections of myth, legend, and esoteric lore. According to her, the Arthurian myth evolved to transform the round zodiac into the Round Table. A construction date of 2700 B.C. was assigned to the Glastonbury Zodiac on the grounds that Taurus and Scorpius fall on the east–west line and so position the equinoxes within them. This orientation would correspond to the equinoxes of the third millennium B.C. This early date has prompted others to suppose that Britain is linked, through the Druid priesthood, to the Sumerian-Chaldean astronomical tradition. The name Somerset is derived from "Summerland," which in turn derives, according to this argument, from "Sumer-land" and refers to an ancient group of settlers from the Middle East who brought their culture to Britain many thousands of years ago.

Even if we assume that the Glastonbury Zodiac figures are not drawn arbitrarily (and this assumption is not necessarily warranted, for Mrs. Maltwood discarded some features and included other more recent components of the landscape to obtain her final versions of the Giants), the effigies do not represent the conventional symbols of the zodiac very well. Aquarius, for example, is present as a phoenix, or, alternatively, an eagle. The two fishes of Pisces are dominated by a third figure, many times larger than either of them. This is said to be the constellation Cetus, the Whale, but it is located in what would be the north part of the sky, when in fact it should be in the south. Libra, the Scales, is somehow miraculously transformed into a dove and is completely mislocated. According to Mrs. Maltwood, Libra was not really included, and the dove is a replacement "air sign" that roughly coincides with Ursa Major and is not in the zodiac at all. Cancer, the Crab, is completely missing, as is Gemini, the Twins. In place of Gemini, the constellations Orion and Argo are supposedly present. Some have identified Mrs. Maltwood's Orion and its twin hills of Lollover and Dundon as Gemini.

It may yet be shown that initiatory rites of some ancient faith in Britain were connected with Glastonbury Tor, the flat-topped natural hill that protrudes above Glastonbury and is now topped by the tower ruins of St. Michael's church. Perhaps some of the features in the Glastonbury landscape were artificially constructed. It is hard to imagine, however, anyone moving enough earth to construct a Leo, the Lion, three miles long. More to the point, most of the zodiac figures are based upon modern field boundaries, and most of these first appeared between 1750 and the time of publication of Maltwood's book. It should be possible to check the Maltwood figures against early estate maps and tithe maps to see if the same figures were present at an earlier date. Glastonbury Abbey carried out extensive drainage programs during the Middle Ages, and this activity must have altered the original landscape in the vicinity. The monuments, if real and if of great antiquity, would probably not be recognizable today. If on the other hand they were designed and constructed in more recent times, say, between the drainage and the time of Dr. Dee, there should be a clearer record of their history.

Since the "discovery" of the Glastonbury Zodiac, several other terrestrial zodiacs have been found, at Nuthampstead in Cambridgeshire, Pumpsaint in Wales, and Kingston-on-Thames in Surrey. Still others have been claimed but have not yet been fully described. All share the same general attributes: truly monumental earthworks whose design and significance can only be appreciated from the air. The familiar case for the existence of ancient astronauts has found its way, finally, to Britain.

What will we accept for proof that these incredibly large figures were sculpted from the earth and not just crafted in the mind's eye? A comparison to the huge pictures and geometric designs drawn on the ground by the so-called Nazca people, a pre-Inca culture that inhabited the coastal valleys and plains near Nazca, Peru, comes to mind, but it would not be just. Archaeologists know that the Nazca lines and figures *were* drawn by an ancient people. They are datable and consistent with the other iconography of the Nazca culture. The construction technique is simple, but the meaning is still a mystery. At Glastonbury it is a mystery whether the figures are really there or are the products of an imaginative mind.

Britain is a long-domesticated landscape with so many layers of inhabitants that it looks to us now, after at least six thousand years of organized activity, that there is meaning in the overlapping fabrics of cultures. That lovely pattern in the crazy quilt may be contrived, however, of whole cloth.

A SERIOUS MYSTERY

Sirius, as we know, is the brightest star of the night sky. Its appearance in the early evening is a sure sign that winter is on the way. The brightness of Sirius might alone have given it special significance in the star lore of ancient peoples, but the coincidence of the heliacal rising of Sirius with the summer solstice and the inundation of the Nile prompted the Egyptians to base an entire calendar and timekeeping system on the behavior of this star. Sir Norman Lockyer argued that seven Egyptian temples, including two at Karnak, had been aligned on Sirius.

The Egyptians were not the only ones to make use of the heliacal rising of Sirius. John Eddy's careful work on the Big Horn Medicine Wheel in Wyoming shows that the North American Indians who built this structure also observed the reappearance of Sirius in the predawn sky.

As distances between the stars go, Sirius is relatively nearby. It is the sixth closest star to earth, at a distance of 2.67 parsecs, or 8.7 light years (approximately 52 trillion miles). Sirius emits considerably more energy than the sun and so, on an absolute scale, is much brighter and hotter than the sun. The surface temperature of Sirius is about 17,500 degrees F., and the star is white in color. It is about twice as massive as our sun.

In 1844 Friedrich Bessel, a German astronomer, concluded that an unseen, dark companion must be in mutual orbit with Sirius. Bessel had completed extremely careful measurements of the slowly changing position of Sirius and had discovered a small wiggle, first to one side and then to other, around the expected line of motion of Sirius in space. Bessel rightly assumed that Sirius was affected by the gravitational influence of a close neighbor, and in 1862, the American telescope maker Alvan Clark saw the faint companion of Sirius for the first time through one of his new instruments.

Sirius A and its companion, Sirius B, revolve about each other in 49.9 years. The companion, it turns out, is a rather exotic type of star that's called a "white dwarf." Although Sirius B contains about as much material as the sun, it is, at most, only about twice as large as the earth. The star is extremely condensed. One teaspoonful of its material weighs nearly a quarter of a ton (which would be rather hard on the teaspoon).

A curious tale about Sirius and its dense, faint companion emerged from Africa a few decades ago. The Dogon, a tribe living in what is now the Republic of Mali, attach special importance to Sirius. In their tradition, the bright Sirius is accompanied by a very heavy and dark object they call "Po," the word the Dogon also use for a genus of cereal grain known as *Digitaria*. This object, according to the Dogon, is the smallest type of star. They also maintain that it completes an orbit around Sirius every fifty years. These beliefs of the Dogon, which were first communicated by tribal priests to two French anthropologists, M. Griaule and G. Deiterlen, in the 1940s, are allegedly at least eight hundred years old, as are the designs of the Dogon sand diagrams that illustrate the Sirius system and other astronomical objects.

One of the Dogon sand diagrams shows what is identified as the elliptical path of Sirius B about Sirius A, with Sirius A located off-center, as though at a focus of the ellipse. The similarity between the Dogon design and the projected relative orbit as seen from the earth is suggestive. The Dogon also hold that Jupiter has four moons and that Saturn is ringed. A telescope is required to see Jupiter's four Galilean satellites and Saturn's rings, but the Dogon have not had the benefit of this instrument. Other traditions of theirs about these planets also predate the telescope's invention.

The startling agreement between modern astronomical knowledge and Dogon star lore prompted an American orientalist, Robert K. G. Temple, to explore the problem in more detail. In his book *The Sirius Mystery*, he details many more Dogon beliefs and aspects of many other mythologies that, in his view, relate to Sirius and its "dark" partner. Temple concludes, in all seriousness, that the original stimulus of Dogon beliefs, and indeed of all of these peculiar beliefs, about Sirius

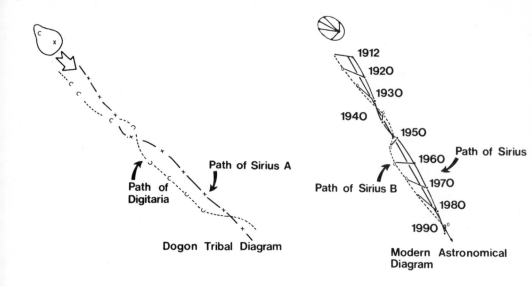

Path of Sirius A

Path of Digitaria

Dogon Tribal Diagram

1912
1920
1930
1940
1950
1960
1970
1980
1990

Path of Sirius

Path of Sirius B

Modern Astronomical Diagram

Robert Temple, in The Sirius Mystery, *illustrates his comparison between Dogon traditions of Sirius and modern astronomical knowledge with these diagrams of the orbit of the companion of Sirius about Sirius itself and the wiggling paths of proper motion, observed and inferred. Actually, this comparison is misleading, for the Dogon do not draw any diagram like that shown at left. The diagram at left is a redrawing by Griffith Observatory, which is based on a reinterpretation of the diagram drawn by Temple. (Griffith Observatory)*

was a visit to earth by extraterrestrial amphibians nearly five thousand or more years ago. According to Temple, the home world of these beings is a planet in the Sirius system.

Temple's case rests upon extensive reinterpretation of a great volume of ancient commentary and mythology. For example, the peculiar myth of Oannes is used to support the idea of planet-hopping amphibians. Oannes, as reported by Berossus, a Babylonian living in the third century B.C., was a peculiar combination of human and fish. Most of Oannes' body was fishlike, but under his fish head was another head. This strange creature, with both a fish's tail and human feet, could speak. It was his custom at evening to return to the sea to spend the night. Other sources characterize Oannes as a teacher who brought writing and astronomy to humanity. Other arts and sciences, in fact, the fundamentals of civilization—agriculture, architecture, law, and mathematics—were all allegedly transmitted to undeveloped cultures of the earth by Oannes.

The myth of Oannes is probably much older than the earliest account (Berossus') of it. Temple is impressed by the fishlike attributes of this culture bearer, and he concludes that Oannes must have come from an oceanic world far from the earth. Quite arbitrarily Temple assumes that the seemingly detailed astronomical knowledge of Sirius held by the Dogon is evidence that Oannes' home world was part of the Sirius system.

Temple believes the Dogon could only have obtained correct detailed knowledge of the Sirius system either with modern telescopes, which of course they did not have, or through direct communication with visiting spacemen. It is difficult, he argues, to believe that the Dogon would have stumbled upon the fifty-year orbital period of Sirius B by coincidence.

Many examples from the myths and legends, including those concerning Gilgamesh, Jason and the Argonauts, the Sumerian Anunnaki, and a monstrous hound of Greek mythology, Orthrus, are associated one way or another with the number 50, and Temple maintains that these elements of the various stories refer to the binary orbit of Sirius.

Dogon astronomical traditions were bound to cause consternation among astronomers, for it seems impossible to reconcile the Dogon astronomical knowledge of the star Sirius with their instrumental resources, namely the unaided eye. Temple's conclusions, in turn, are so unorthodox that his evidence and his handling of it demand the most careful review.

It does seem remarkable that a tribe in Africa, which is celebrated for its complex system of cosmology and unique religious traditions, should have detailed knowledge of the existence and nature of Sirius B, the Galilean satellites of Jupiter, and the rings of Saturn. The possibility always remains that the apparent similarity between the Dogon description of the Sirius system and the real situation is due to chance. Some may be as uncomfortable, however, with this explanation as with Temple's extraterrestrial hypothesis. We may never know for certain why the similarities exist, but it is possible to remain skeptical of the notion of an ancient visit to earth by amphibious inhabitants of Sirius.

It should not surprise us, on the other hand, that Sirius occupies a special place in the Dogon scheme, since it is the brightest star in the sky. It was, as we have learned, a convenient reference for the Egyptian solar calendar, and although we do not know if Egyptian tradition was transmitted to the Dogon, the apparent brightness of Sirius could have stimulated an independent Dogon tradition.

Temple rightly cautions us to be wary of accepted translations of ancient texts and interpretations of myths, yet he seems to fail to heed his own advice. His catalogue of the occurrence of the number 50 is a case in point. Associations like these are always suggestive, but they are al-most invariably taken out of context, in Velikovskian fashion. Perhaps we should attach special significance to these relationships, but we have no way to judge unless all such numerical associations are tabulated and classified. We may be finding 50s in myths because we are looking for 50s. An objective test should be devised if we are going to imagine that we are doing science.

The appearance of Oannes is another troublesome matter in Mesopo-tamian tradition, but alternative explanations to the "visitor-from-Sirius" hypothesis are available. Oannes may be associated with the Sumerian god Ea, who also provided humanity with wisdom and the arts of civilization. According to Berossus, Oannes came from the waters of the Persian Gulf, and this may refer to a seafaring people whose major center was located on the island of Bahrain. It is thought by some archaeologists that these maritime people did introduce new culture to the "land between the rivers" and stimulated the growth of Sumerian civilization.

There is no direct connection between Oannes and the star Sirius, but we might reasonably insist upon one if we are to believe Temple's notion that the myth of Oannes echoes the visit to earth of amphibious creatures from a watery planet in the Sirius system. Otherwise, we might arbitrarily choose any other god of the ancient Near East and deduce from his or her character the physical conditions upon a hypo-thetical planet from whence the god had come. Thoth, the Egyptian ibis-headed god of science, learning, and writing, might suggest equally well that a planet of the Sirius system is a vast marshland.

Lockyer, in *The Dawn of Astronomy*, associated Oannes with Ea and reminded us that Professor Peter Jensen, a German orientalist and au-thor of *Cosmologie der Babylonier* (1890), thought the star Eta Argus (now Eta Carinae) was Ea's star. This peculiar eruptive variable is known for its irregular variations in light. Perhaps the great brightness of Sirius was misequated with the past great brightness of Eta Carinae, which at times outshone even Canopus, or Alpha Carinae, the brightest star after Sirius. The myth of the dark companion may be a recollection of the erratic fading of Eta Carinae.

Perhaps the strongest evidence against the idea that the Dogon tradi-tion is a relic of extraterrestrial contact is found among the Dogon themselves. Not only do the Dogon speak of a second, dark member of the Sirius system, but they also describe a third star, "Emme Ya," and the planet that orbits it. Emme Ya requires thirty-two years to com-plete an orbit. Though only one fourth as massive as Sirius B, it is larger than that famous white dwarf. There is no observational evidence that yet another companion orbits Sirius (with or without planet). If the Dogon tradition of a dark neighbor of Sirius truly refers to Sirius B, what are we to make of this additional partner, which apparently does not exist? Furthermore, why are the orbital periods of these stars preserved in years, a strictly earthbound unit of time, as opposed to some untranslatable unit of time that would have been significant only to inhabitants of Sirius. It also seems odd that visitors from Sirius would provide the Dogon with a plot of the relative orbit of Sirius B as seen from the earth. It is more likely this Dogon diagram somehow derives from relatively recent earth-based astronomy.

Perhaps the most revealing item of all is the Dogon statement that Jupiter has four satellites. At first, this claim seems amazing, for Galileo did not know Jupiter had any satellites until he viewed four of them through his telescope. Without telescopes of their own, the Dogon must surely have been given the information by someone else. Whoever the source of this information may have been, it is certain ancient astronauts did not provide it. Jupiter is now known to have at least fourteen satellites, and visiting spacemen inclined to pass this kind of information along should have been more accurate.

Although we may not be able to identify the source of the Dogon Sirius mystery, it seems more likely that their astronomical ideas are either a collection of good and bad guesses or a garbled record of old and recent astronomical knowledge which somehow contaminated older Dogon beliefs.

ASSORTED ANCIENT ASTRONAUTS

Erich von Däniken, a Swiss innkeeper-turned-cosmologist, is a force to be reckoned with. His four books on ancient astronauts have sold in the tens of millions, world-wide in many languages, and that figure alone would classify his work as astronomical. Two motion pictures and two television specials have been inspired by his books. Nearly four dozen similar books, by a variety of authors, have been spawned by his ideas. Doubtless more are on the way. The "chariots of the gods" have been panned in a magazine parody, "Hot Rods of the Gods," and have been propelled into at least three comic book versions of the theme. A juvenile edition of *Chariots of the Gods?* is available in the children's departments of better bookshops. Von Däniken has been interviewed in the United States in every medium, from Johnny Carson's TV show to *Playboy* magazine. He is clearly not going to go away. He has touched something in our imaginations and in doing so has rattled loose some of the underpinnings of our intellects. He has revitalized P. T. Barnum's assessment of the skepticism and intelligence of the human mind, and astronomy and archaeology are left holding the bag.

In Von Däniken's books artifacts from widely separated times and places are assembled together. By their mysterious forms and allegedly unknown meanings they are alleged to compel us to believe that visitors from space came to earth in ancient times. These ancient astronauts were viewed by primitive earthmen as gods. Such a spaceman is a literal *deus ex machina* that transforms human genetics, culture, and history at once.

Von Däniken repeatedly misinterprets the meaning of many astronomically, related artifacts and, worse yet, shortchanges the powers of observation and knowledge of ancient peoples. His exploitation of the Maya calendar is typical.

The Maya calendar is famous for its great accuracy. This achievement, combined with the enigmatic appearance and decline of Maya civilization, is too great a temptation for Von Däniken. He quotes the Maya determination of the length of the tropical year, 365.2420 days, and comments that this is amazingly close to the true length, only 0.0002 day off. This marvelous success does not square with Von Däniken's description of the Maya. He calls them a "jungle people," and he cannot accept that a jungle people could achieve such a result independently. They must have had outside help. Von Däniken's characterization of the Maya is not only inaccurate; his argument is also blatantly racist.

More suggestive arguments from the Maya are invoked to support Von Däniken's picture of astronaut intervention. Von Däniken repeatedly refers to the "incredible Venus formula," a calendar cycle of 37,960 days that brought the sacred day count (which Von Däniken mistakenly associates with the moon), the tropical year, and the synodic period of Venus into commensurability. What is most incredible about the Venus formula is that Von Däniken mistakes the synodic period of Venus for its true orbital period and then implies that the Maya must have had privileged information from astronauts who observed Venus from space. Here Von Däniken's ignorance of astronomical terms is compounded by his disregard for the fact that we have figured out the sidereal period of Venus without actually traveling there.

A really puzzling detail of Von Däniken's commentary on the Maya calendar is his inclusion of subscripts in the specification of the number of days in certain intervals of time. For example,

"20 kinchiltuns$=$1 atautun or 23$_2$,040$_1$,000,000 days"

The conscientious student of archaeoastronomy might well ask what the meaning of those curious subscripts is and how it could be that they have been overlooked in the past. In fact, the subscripts are nonsense, have nothing to do with the Maya calendar, and have no business in the discussion at all. What Von Däniken means by them remains a mystery.

Von Däniken also implies that the Maya time cycles were used to predict the return of the astronaut-gods. After a length of time the gods "would come to the resting place," he says. In reality, the "gods" were the days themselves, who carried their burdens of time to the periodic return, or resting place, of the calendar round. The resting places are just the dates when certain astronomical cycles were completed.

It is difficult to chronicle all of Von Däniken's astronomical illiteracy. On the one hand he is nonsensical. His statement, "The Maya astronomers knew the moon's orbit to four decimal places . . ." is meaningless. On the other hand, he is ludicrous. Von Däniken asks, "How did the Mayas know about Uranus and Neptune?" There is no evidence that they did. And if they did, why didn't they know about Pluto?

The Caracol at Chichén Itzá has been proposed as being a Maya observatory by several investigators, although they do not agree on which astronomical alignments were built into this structure. Von Däniken enthusiastically agrees that the Caracol must be an observatory because it looks like an observatory. It has a dome and provides, he tells us, "an impressive picture of the firmament at night." Actually, the Caracol's now ruined "dome" was tower-shaped, and if it provides an impressive picture of the firmament at night, it is because half of the roof is caved in. The view from the upper story was limited to a few precisely placed windows. Von Däniken asks, quite mysteriously, why the windows do not point to the brightest stars. Perhaps it is because the sun and Venus may have been intended targets.

Von Däniken also reveals that archaeologists have not been completely truthful in their reports on the sacred well, or *cenote*, at Chichén Itzá. This well, he says, fits the scholars' theories, but there is a second well, a secret, hidden well, Von Däniken says, that has been intentionally ignored by the scientists because they can't explain it. Well, Von Däniken is right: there *is* a second well at Chichén Itzá, and a third well and a fourth well—over twenty altogether. But all of these have been documented by the archaeologists, including the "second secret, hidden well," which is located right in the middle of Chichén Itzá and is pretty hard to miss if you're looking for it, and the wells do fit the archaeologists' theory of their use, namely, the Maya needed water. The two main wells of Chichén Itzá were "possibly formed by the impact of meteorites," according to Von Däniken. This suggestion is also nonsense, and it is geologically obvious that the wells are limestone sinkholes.

One more Maya artifact is worth considering, the sarcophagus cover from the tomb inside the Temple of the Inscriptions at Palenque, in south Mexico. Von Däniken regards this as one of the most convincing pieces of evidence in his file, although in his books and interviews he has trouble keeping straight where both the temple and even Palenque are located.

The Palenque sarcophagus depicts a reclining figure that Von Däniken insists is an astronaut. All of the curvilinear paraphernalia around him are allegedly parts of a rocket ship, in which the astronaut is seated and appropriately attired for blastoff. Von Däniken's space jacket is actually a pair of wrist bracelets and a bare chest. The tight-fitting trousers are so tight as to be invisible. The helmet is really a hairdo, but Von Däniken's picture of it is sufficiently smudged to prevent the reader from seeing the details. The astronaut's padded seat is an altar. This altar sits on top of Von Däniken's rocket exhaust, which is a very obvious earth monster. The whole affair is contained within the stylized jaws of a reptile, another typical motif, and from the astronaut, stretching upward is the rocket's fuselage, in reality, the Maya Tree of Life. Perched at the top of the rocket is a quetzal bird, presumably a hood ornament. All of these Maya symbols are found in many other Maya sculptures. Even at Palenque other variations have been found in which the astronaut is absent but in which priests are shown in attendance at the altar. These images support the altar interpretation and weaken Von Däniken's claim.

Because the pre-Columbian civilizations of South America are not well known to most of us, Von Däniken is able to take what he wants from them with little regard for archaeological truth. More than twelve thousand feet above sea level on the Bolivian plateau, near Lake Titicaca, is Tiahuanaco, a site to which Von Däniken is compelled to transport us again and again.

Like so many of the legitimate archaeological mysteries that Von Däniken plunders, Tiahuanaco is ripe for exploitation. There is much about it we do not know. It is not regarded as an ancient metropolis but as a pre-Inca ceremonial center. It was built and used by a people who invaded Peru and overran the Nazca and Mochica cultures that preceded them. In *Chariots of the Gods?* Von Däniken accepts the altogether fanciful date of 27,000 B.C. for the origin of Tiahuanaco. In a more recent book, *In Search of Ancient Gods*, he is just as happy to use the date A.D. 600 (supplied by the archaeologists he is so ready to criticize), because it usefully coincides with his dates for biblical astronauts.

Von Däniken appears to have excerpted most of his claims about Tiahuanaco from *The Morning of the Magicians* by Louis Pauwels and Jacques Bergier. Their book is a massive collection of historical oddities, unsubstantiated rumors, secret knowledge, occult lore, and random science designed to convince us that the world is not as it seems. Fair enough, but *The Morning of the Magicians* is every bit as undisciplined as Von Däniken's books and certainly no more reliable.

Pauwels and Bergier in turn rely upon a book by Hans Bellamy and Peter Allen, *The Great Idol of Tiahuanaco*. According to the authors, the Great Idol's symbols and inscriptions contain considerable astronomical and calendrical data. These data, quite amazingly, are consistent with Hanns Hörbiger's theories of the solar system's early history. On this unlikely result rests Von Däniken's claim that the Tiahuanacan knowledge of astronomy was highly sophisticated, too sophisticated for those primitive folk, and of great antiquity.

The reader is not informed, however, that Bellamy and Allen are not specialists in pre-Columbian iconography and that no recognized scholar has verified their "translations." Nor are we told that Bellamy wrote a number of books in the 1930s, among them *Moons, Myths, and Man* (1936), which supported and publicized the theories of Hanns Hörbiger in England and America. In the 1920s Hörbiger treated the world to his "cosmic ice theory," in which the moon and all the planets have thick outer layers of ice and the Milky Way is just a ring of orbiting ice blocks. It is not surprising Bellamy would find confirmation of Hörbiger twenty years later at Tiahuanaco.

Von Däniken also disregards the fact that Hörbiger's theories are astronomical nonsense. In a pre-Velikovskian series of cataclysms, according to Hörbiger, the earth captured and then lost at least six moons. We are in the age of the seventh moon now. The decoding of Tiahuanaco by Bellamy and Allen showed that it was built in the age of the sixth moon and records astronomical data consistent with that era. If we believe Bellamy and Allen, then we must conclude either poor astronomy was practiced at Tiahuanaco by the natives or that the visiting spacemen weren't very good at it either.

One of the strongest images for those who have seen the motion picture or television presentations of Von Däniken's "chariots" theme is the pattern of lines and figures on the Nazca plain of Peru. Again, Von Däniken suggests that a monument little known to the average person has been ignored by scholars. He accuses scholars with failing to understand how the figures were inscribed, and for what purpose, and with neglecting them to avoid professional embarrassment.

Von Däniken comes to our rescue, however, with the obvious solution that these signs were indicators and landing strips laid out by astronauts for an interstellar airport. He insists that his must be the correct solution because no primitive people could carry out such a task on their own. He adds that certain lines are aligned with other significant sites hundreds of miles away.

There are so many lines at Nazca, however, that we would expect statistically that some of them would point to special places, perhaps even Washington, D.C., and Niagara Falls. We shall count ourselves lucky, in fact, if one doesn't point to the Great Pyramid in Egypt. A few alignments in themselves do not prove that alignments were intended. Alignments must be classified and their precisions specified before a statistical test can be performed.

It is not true, also, that the lines and drawings must have been laid out and seen only from above. Relatively simple surveying and mural cartoon techniques would have permitted construction of the forms. And there are nearby mountains from which some of the designs could have been viewed.

It is hard to believe that visiting spacemen, who must have the technological capacity to travel hundreds of light-years to earth, would require either landing strips or gigantic navigational markers once they arrived. The United States Apollo astronauts landed safely upon the moon in 1969 without arrows and landing circles etched onto the surface.

Some are too quickly tempted to accept Von Däniken's claim that the whole Nazca area looks like an airport. His books never show the entire region, roughly sixty miles by ten miles, in which lines of varying width, triangles, and pictures of animals and plants crisscross the flatland and hills. Certain sections of the area bear a superficial resemblance to an airport, a resemblance that dissolves when one looks again at aerial views of airports and is startled by their relative simplicity. We don't find monkeys and spiders inscribed across our runways. These animal figures and geometric patterns, created on the ground by simply moving the surface rocks aside, are duplicated on Nazca pottery. It is clear that the Nazca ground drawings were done by the same pre-Inca people who made the pottery and were not the product of some much earlier and unknown race, as Von Däniken implies.

Among the many Egyptian mysteries Von Däniken discusses, at least one is connected with astronomy. The Egyptian calendar, which was divided into three seasons determined by the agricultural cycle and the behavior of the Nile River, was calibrated by the predawn rising of Sirius. As we have already seen, Sirius first reappeared to rise heliacally at the time the Nile overflowed its banks. Von Däniken is unsatisfied with the calendrical explanations of Sirius and suggests that the Egyptians had ulterior motives for placing such emphasis upon it. Sirius, he suggests, may have been the home of astronaut gods. To support this notion, he points out that the Nile flooding did not occur on the same day every year, and so Sirius would have been useless for this purpose. This claim is ridiculous considering that we, who use a tropical calendar based on the equinoxes and solstices, do not expect the snow to fall or the leaves to turn on the first days of winter and fall.

Von Däniken asks, "Is it a coincidence that the area of the base of the pyramid divided by twice its height gives the celebrated figure $\pi=3.14159$?" No, because it doesn't, and even if it did it would be a coincidence because the result of the calculation is strictly dependent upon the units chosen. Here we are dividing an area by a length, and any answer desired is possible. Von Däniken meant to ask about the much-touted ratio of the length of the perimeter around the base to the height of the pyramid, and a variety of sensible explanations have been available for that result.

Von Däniken asks, "Is it a coincidence that the meridian running through the pyramids divides the continents and the oceans into exactly two equal halves?" Yes, it probably is, but we really don't know, because whatever Von Däniken means by "two exactly equal halves" is nonsense.

Von Däniken asks, "Is it really a coincidence that the height of the pyramid of Cheops multiplied by a thousand million—98,000,000 miles —corresponds approximately to the distance between earth and sun?" Since the ancient astronauts who provided this figure must have known that this estimate is 5.4 million miles off, it must be a coincidence. Worse yet, Von Däniken has both used an incorrect height for the pyramid and made a multiplication mistake. Even doing it right, the distance to the sun is not correctly obtained by the calculation. Most everything Von Däniken has to say about ancient Egypt is of comparable accuracy.

Von Däniken feels he is able to pinpoint the times of contact between extraterrestrials and earthfolk to distinct periods: one very early, between 25,000 and 40,000 B.C.; a second perhaps between 7000 and 3500 B.C.; and finally, a third visit, in 592 B.C. This last landing coincides with the vision reported by the Old Testament prophet Ezekiel. Ezekiel's description of the fiery wheel is regarded by Von Däniken as a "showpiece" in his chain of evidence. The cloud, the fire, the four shapes with the likenesses of living creatures, the wheels within wheels, and all of the other exotic details are taken to be a literal description of the landing of a space vehicle. Josef Blumrich, chief of systems layout of NASA, undertook a serious examination of Ezekiel's vision and was able to extract from it a very plausible spacecraft. His efforts are described in his own book, *The Spaceships of Ezekiel.*

Interesting as Blumrich's conclusions are, their validity is compromised by an incomplete understanding of the symbols. The landing legs on Blumrich's vehicle are the literal counterparts of Ezekiel's creatures, and Ezekiel tells us they each bear four faces, of an ox, a man, a lion, and an eagle. Literal interpretation of Ezekiel requires that these details fit into the over-all scheme, and Blumrich, quite imaginatively, makes them out to be descriptions of protective fairings and cutouts on the landers. Normally, if no known alternative is available to counter Von Däniken, his own enthusiasm carries Von Däniken along. Here, at least, skeptics are on firm ground. The four animals mentioned have long been known as the four cherubim. Originally they were temple guardians used by peoples in Western Asia, and later they were adopted as the symbols of the four Christian Evangelists. They possibly correspond to, and possibly originated from, the four constellations of the zodiac in which the solstices and equinoxes were located when the vernal equinox was in Taurus, the Bull. The man equates with Aquarius, as does the lion with Leo. The eagle may equate with Scorpius, which according to some commentators was regarded by some Mesopotamian peoples as an eagle and not as a scorpion. Perhaps these astronomical interpretations are themselves incorrect, but the similarity between them and the figures described in Ezekiel suggests a more plausible line of further inquiry.

Von Däniken urges us to marvel that the solar and lunar eclipses are predicted on a Babylonian tablet in the British Museum, though he does not check to see if the predictions are correct. Moreover, we have recognized for a long time that the Babylonians had the potential to make some eclipse predictions. It would be even more amazing if cosmic visitors to earth could figure out all of this information, or would even bother to, during what must have been a brief stay.

Much more sophisticated astronomical knowledge is found by Von Däniken in the apocryphal Book of Enoch. He astounds his readers with the knowledge that Chapters 72–82 give "accurate details of the orbits of sun and moon, intercalary days, the stars and the functioning of the heavens." Von Däniken repeatedly confuses a planet's true (sidereal) orbit and its periodic motion as seen from the earth (synodic period) and permits us to conclude he is again referring to basic observational data. These data can be acquired without telescopes, and calculations can be made without assuming or knowing that the earth and other planets revolve around the sun. Von Däniken, nevertheless, is in the dark. He equates Enoch's simple description of the existence of the ecliptic with Johannes Kepler's three laws of planetary motion and misleads the reader into believing Enoch's astronomy is more sophisticated than it really is. Enoch goes on to say that the stars of heaven were all named by their names and weighed with a genuine measure, according to the intensity of their light. But these are things modern astronomers do, Von Däniken points out, so Enoch could hardly have known about them without astronaut assistance. Obviously Von Däniken's astronomical naivety is equaled only by his ignorance of history.

One picture is said to be worth a thousand words, but it is hard to assess the value of either words or pictures if it's Von Däniken's that we're counting. The pictures in *In Search of Ancient Gods*, subtitled *My Pictorial Evidence for the Impossible*, are a welcome relief from Von Däniken's verbiage, but they show again how fast and loose he plays the game. In an "extraordinary rock drawing" from a cave in Brazil, the caption adds, "Eight of the nine planets are drawn in their correct relations to the sun—proof of astronomical knowledge that Stone Age artists just did not have. Who were their teachers?" Von Däniken does not explain what he means by "Stone Age" nor by what method the drawing was dated nor why Pluto is missing. Von Däniken accompanies the photograph with a diagram of the drawing on which the planets are properly identified. A comparison of the diagram and photograph show there is little resemblance between them, and in Von Däniken's interpretive diagram Mars appears to be more distant from the sun than Jupiter. Also, Uranus is placed more distant from the sun than Neptune. Neither distances nor diameters are properly scaled, and the number of satellites in almost every case is wrong.

Apart from any criticism we might make of Von Däniken's handling and mishandling of the data, we can still ask whether on his own terms he proves his case. One way of evaluating him is the testing of his main point, that earth was visited on two or three occasions in the past, during the periods noted earlier. If we plot Von Däniken's evidence on a time line, we expect to see the points clustered around the dates he indicates that we were visited by astronauts. Only the artifacts mentioned by Von Däniken are included, and his dates for them, however outrageous, are used when, infrequently, he gives them. Otherwise the best available dates are used. From the diagram we can see that even on his own terms, Von Däniken has failed to prove his point. Far from recording a few isolated visits, the diagram suggests the existence of a regularly scheduled shuttle between earth and the stars. Perhaps the earth is really the "vacation spa of the gods." If anything, traffic in extraterrestrial astronauts has increased in the last two thousand years.

Perhaps we should not be so hard-nosed. It would be very difficult to prove that ancient astronauts did not visit the earth. But evidence for their having been here is just not convincing.

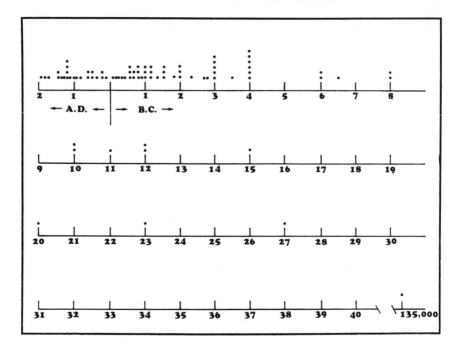

The distribution in time of Von Däniken's archaeological mysteries contradicts his own hypothesis of when the ancient astronauts came to earth. If anything, it appears as though their visits have continued more frequently during the last two thousand years. Each mark on the line stands for one thousand years. The date Von Däniken assigns to each artifact and piece of evidence, no matter how outrageous, is plotted for each artifact for which he provides a date. Otherwise, the most scientifically reliable date is used. (E. C. Krupp)

Somewhere in this dreary exposé of human gullibility something good may yet emerge. Many people are now aware of a rich heritage of human past that has gone relatively unnoticed, save for the efforts of a few dedicated scholars. Modern science might well learn a lesson from the Von Däniken phenomenon. Von Däniken gave his audience what it wanted: a renewed sense of the mystery and wonder of the cosmos, of the earth, and of human history. If scientists can learn to communicate their discoveries with the same enthusiasm as Von Däniken does and if we can learn to partake of the tree of knowledge without losing the innocence and curiosity that prompted our first questions, the human spirit will continue to evolve.

Bibliography

CHAPTER 2

BROADBENT, S. R. "Quantum Hypotheses," *Biometrika*, vol. 42 (1955), pp. 45–57.
————. "Examination of a Quantum Hypothesis Based on a Single Set of Data," *Biometrika*, vol. 43 (1956), pp. 32–44.
HUTCHINSON, G. E. "Long Meg Reconsidered," *American Scientist*, vol. 60 (1972), pp. 24–31.
————. "Long Meg Reconsidered, Part 2," *American Scientist*, vol. 60 (1972), pp. 210–19.
KENDALL, D. G. "Hunting Quanta," *Philosophical Transactions of the Royal Society of London*, vol. 276 ("The Place of Astronomy in the Ancient World," ed. F. R. Hodson), 1974, pp. 231–66.
MACKIE, E. W. "Archaeological Tests on Supposed Prehistoric Astronomical Sites in Scotland," *Philosophical Transactions of the Royal Society of London*, vol. 276 ("The Place of Astronomy in the Ancient World," ed. F. R. Hodson), 1974, pp. 169–94.
ROY, A. E., et al. "New Survey of the Tormore Circles (Arran)," *Transactions of the Glasgow Archaeological Society*, vol. 15 (1963), pp. 59–67.
THOM, A. "The Solar Observatories of Megalithic Man," *Journal of the British Astronomical Association*, vol. 64 (1954), pp. 396–404.
————. "A Statistical Examination of the Megalithic Sites in Britain," *Journal of the Royal Statistical Society*, vol. 118 (1955), pp. 275–95.
————. "The Egg-Shaped Standing Stone Rings of Britain," *Archives Internationales d'Histoire des Sciences*, vol. 14 (1961), pp. 291–303.
————. "The Geometry of Megalithic Man," *Mathematical Gazette*, vol. 45 (1961), pp. 83–93.
————. "The Megalithic Unit of Length," *Journal of the Royal Statistical Society*, vol. 125 (1962), pp. 243–51.
————. "The Larger Units of Length of Megalithic Man," *Journal of the Royal Statistical Society*, vol. 127 (1964), pp. 527–33.
————. "Megalithic Geometry in Standing Stones," *New Scientist*, March 12, 1964.
————. "Observatories in Ancient Britain," *New Scientist*, July 2, 1964.
————. "Megalithic Astronomy: Indications in Standing Stones," in A. Beer, ed., *Vistas in Astronomy*, vol. 7. Oxford: Pergamon Press, 1966. Pp. 1–57.
————. "Megaliths and Mathematics," *Antiquity*, vol. 40 (1966), pp. 121–28.
————. *Megalithic Sites in Britain*. London: Oxford University Press, 1967.
————. "The Metrology and Geometry of Cup and Ring Marks," *Systematics*, vol. 6 (1968), pp. 173–89.
————. "The Geometry of Cup and Ring Marks," *Transactions of the Ancient Monuments Society*, vol. 16 (1969), pp. 77–87.
————. "The Lunar Observatories of Megalithic Man," in A. Beer, ed., *Vistas in Astronomy*, vol. 11. Oxford: Pergamon Press, 1969. Pp. 1–29.
————. "The Megaliths of Carnac," *The Listener*, vol. 84 (1970), pp. 900–1.
————. "Observing the Moon in Megalithic Times," *Journal of the British Astronomical Association*, vol. 80 (1970), pp. 93–99.
————. *Megalithic Lunar Observatories*. London: Oxford University Press, 1971.
————. "Astronomical Significance of Prehistoric Monuments in Western Europe," *Philosophical Transactions of the Royal Society of London*, vol. 276 ("The Place of Astronomy in the Ancient World," ed. F. R. Hodson), 1974, pp. 149-56.
————. "A Megalithic Observatory in Islay," *Journal for the History of Astronomy*, vol. 5 (1974), pp. 50–51.

————, and A. S. THOM. "The Astronomical Significance of the Large Carnac Menhirs," *Journal for the History of Astronomy*, vol. 2 (1971), pp. 147–60.

————. "The Carnac Alignments," *Journal for the History of Astronomy*, vol. 3 (1972), pp. 11–26.

————. "The Uses of the Alignments at Le Ménec, Carnac," *Journal for the History of Astronomy*, vol. 3 (1972), pp. 151–64.

————. "A Megalithic Lunar Observatory in Orkney," *Journal for the History of Astronomy*, vol. 4 (1973), pp. 111–23.

————. "The Kerlescan Cromlechs," *Journal for the History of Astronomy*, vol. 4 (1973), pp. 168–73.

————. "The Kermario Alignments," *Journal for the History of Astronomy*, vol. 5 (1974), pp. 30–47.

————. "Further Work on the Brogar Lunar Observatory," *Journal for the History of Astronomy*, vol. 6 (1975), pp. 100–14.

————. "Megalithic Astronomy," *The Journal of Navigation*, vol. 30, no. 1 (1977), pp. 1–14.

————. "A Fourth Lunar Foresight for the Brogar Ring," *Journal for the History of Astronomy*, vol. 8 (1977), pp. 54–55.

THOM, A., et al. "The Astronomical Significance of the Crucuno Stone Rectangle," *Current Anthropology*, vol. 14 (1973), pp. 450–54.

————. "Stonehenge," *Journal for the History of Astronomy*, vol. 5 (1974), pp. 71–90.

————. "Stonehenge as a Possible Lunar Observatory," *Journal for the History of Astronomy*, vol. 6 (1975), pp. 19–30.

————. "The Two Megalithic Observatories at Carnac," *Journal for the History of Astronomy*, vol. 7 (1976), pp. 11–26.

————. "Avebury (1): A New Assessment of the Geometry and Metrology of the Ring," *Journal for the History of Astronomy*, vol. 7 (1976), pp. 183–92.

————. "Avebury (2): The West Kennet Avenue," *Journal for the History of Astronomy*, vol. 7 (1976), pp. 193–97.

CHAPTER 3

ATKINSON, R. J. C. *Stonehenge*. London: Hamish Hamilton, 1956.

————. "Moonshine on Stonehenge," *Antiquity*, vol. XL (1966), pp. 212–16.

————. *Stonehenge and Avebury*. London: Her Majesty's Stationery Office, 1971.

————. "Megalithic Astronomy—A Prehistorian's Comments," *Journal for the History of Astronomy*, vol. 6 (1975), pp. 42–53.

————. "The Stonehenge Stations," *Journal for the History of Astronomy*, vol. 7 (1976), pp. 142–44.

————. "Interpreting Stonehenge." *Nature*, vol. 265 (1977), p. 11.

BEACH, A. D. "Stonehenge I and Lunar Dynamics," *Nature*, vol. 265 (1977), pp. 17–21.

BRINCKERHOFF, RICHARD F. "Astronomically-Oriented Markings on Stonehenge," *Nature*, vol. 263 (1976), pp. 465–68.

BROWN, PETER LANCASTER. *Megaliths, Myths and Men*. Poole, Eng.: Blandford Press, 1976.

BUNTON, GEORGE W., and JOSEPH E. CIOTTI. "Stonehenge—a Fifty-Six Year Eclipse Cycle?" *The Griffith Observer* (Los Angeles), vol. 40 (April 1976), no. 4, pp. 7–11.

BURL, AUBREY. *The Stone Circles of the British Isles*. New Haven: Yale University Press, 1976.

CRAMPTON, PATRICK. *Stonehenge of the Kings*. London: John Baker Publishers, 1967.

FERGUSSON, JAMES. *Rude Stone Monuments*. London: John Murray, 1872.

GEOFFREY OF MONMOUTH. *The History of the Kings of Britain.* Tr. Lewis Thorpe. Harmondsworth, Eng.: Penguin Books, 1966.

FORDE-JOHNSTON, J. *Prehistoric Britain and Ireland.* New York: W. W. Norton & Company, 1976.

HADINGHAM, EVAN. *Circles and Standing Stones.* New York: Walker and Company, 1975.

HARRISON, W. J. "Bibliography of Stonehenge and Avebury," *Wiltshire Archaeological Magazine,* vol. XXXII (1902), pp. 1–169.

HAWKINS, GERALD S. *Stonehenge Decoded.* Garden City, N.Y.: Doubleday & Company, 1965.

———. *Beyond Stonehenge.* New York: Harper & Row, 1973.

HAWKES, JACQUETTA. "God in the Machine," *Antiquity,* vol. XLI (1967), pp. 174–80.

HOYLE, FRED. "Speculations on Stonehenge," *Antiquity,* vol. XL (1966), pp. 262–76.

———. "Hoyle on Stonehenge: Some Comments," *Antiquity,* vol. XLI (1967), pp. 91–98.

———. *From Stonehenge to Modern Cosmology.* San Francisco: W. H. Freeman and Company, 1972.

———. *On Stonehenge.* San Francisco: W. H. Freeman and Company, 1977.

HUTCHINSON, H. N. *Prehistoric Man and Beast.* New York: D. Appleton and Company, 1897.

JONES, INIGO. *Stone-henge* (1655). Menston, Yorkshire: Scolar Press, 1973. Reprint of 1655 edition.

KRUPP, E. C. "Stonehenge: The New Astronomy (a Brief Introduction)," *The Griffith Observer* (Los Angeles), vol. 40 (1976), no. 4, pp. 2–7.

LOCKYER, J. NORMAN. Two reviews of books on Stonehenge, *Nature,* vol. 66 (1902), pp. 25–27.

———. *Stonehenge and Other British Stone Monuments Astronomically Considered.* London: MacMillan & Company, 1909.

———, and F. C. PENROSE. "An Attempt to Ascertain the Date of the Original Construction of Stonehenge from Its Orientation," *Nature,* vol. 65 (1901), pp. 55–57.

MACKIE, E. W. *Science and Society in Prehistoric Britain.* London: Paul Elek, 1977.

MICHELL, JOHN. *A Little History of Astro-archaeology.* London: Thames and Hudson, 1977.

NEWALL, R. S. *Stonehenge Wiltshire (Official Guidebook).* London: Her Majesty's Stationery Office, 1971.

NEWHAM, C. A. *The Enigma of Stonehenge.* Sedge Rise, Tadcaster, Yorkshire: C. A. Newham, 1964.

———. "Stonehenge—a Neolithic Observatory," *Nature,* vol. 211 (1966), pp. 456–58.

———. *Supplement to "The Enigma of Stonehenge."* Leeds, Yorkshire: John Blackburn, 1970.

———. *The Astronomical Significance of Stonehenge.* Leeds, Yorkshire: John Blackburn, 1972.

RENFREW, COLIN. *Before Civilization.* New York: Alfred A. Knopf, 1973.

———. *British Prehistory.* Park Ridge, N.J.: Noyes Press, 1974.

ROBINSON, JACK H. "Sunrise and Moonrise at Stonehenge," *Nature,* vol. 225 (1970), pp. 1236–37, 1970.

STONE, J. F. S. *Wessex Before the Celts.* New York: Frederick A. Praeger, 1960.

THOM, A., et al. "Stonehenge," *Journal for the History of Astronomy,* vol. 5 (1974), pp. 71–90.

———. "Stonehenge as a Possible Lunar Observatory," *Journal for the History of Astronomy,* vol. 6 (1975), pp. 19–30.

WERNICK, ROBERT. *The Monument Builders.* New York: Time-Life Books, 1973.

CHAPTER 4

ALLEN, RICHARD H. *Star Names, Their Lore and Meaning.* New York: Dover Publications, 1963.

BAITY, ELIZABETH C. "Archaeo-Astronomy and Ethno-Astronomy So Far," *Current Anthropology*, 14 (1973), p. 389.

BRANDON, WILLIAM. *The American Heritage Book of Indians.* New York: Dell Publishing Company, 1961.

BRANDT, J. C., et al. "Possible Rock Art Records of the Crab Nebula Supernova in the Western United States," in A. F. Aveni, ed., *Archaeoastronomy in Pre-Columbian America.* Austin: University of Texas Press, 1975. Pp. 45–58.

BRITT, CLAUDE, JR. "Early Navajo Astronomical Pictographs in Canyon de Chelly, Northeastern Arizona, USA," in A. F. Aveni, ed., *Archaeoastronomy in Pre-Columbian America.* Austin: University of Texas Press, 1975. Pp. 89–108.

BROWN, L. A. "The Fort Smith Medicine Wheel, Montana," *Plains Anthropologist*, 8 (1963), no. 22.

CHAMBERLAIN, VON DEL. "American Indian Interest in the Sky Indicated in Legend, Rock Art, Ceremonial and Modern Art," paper presented at AAAS-CONACYT Meeting on Archaeoastronomy in Pre-Columbian America, Mexico City, June 1973.

COWAN, THADDEUS M. "Effigy Mounds and Stellar Representation," in A. F. Aveni, ed., *Archaeoastronomy in Pre-Columbian America.* Austin: University of Texas Press, 1975. Pp. 217–35.

EDDY, JOHN A. "Astronomical Alignment of the Big Horn Medicine Wheel," *Science*, 184 (1974), pp. 1035–43.

ELLIS, FLORENCE H. "A Thousand Years of the Pueblo Sun-Moon-Star Calendar," in A. F. Aveni, ed., *Archaeoastronomy in Pre-Columbian America.* Austin: University of Texas Press, 1975. Pp 59–88.

FEWKES, J. WALTER. "A Sun Temple in the Mesa Verde National Park," *Art and Archaeology*, vol. 3 (1916), pp. 341–46.

FOWLER, MELVIN L. "A Pre-Columbian Urban Center on the Mississippi," *Scientific American*, vol. 233 (August 1975), pp. 92–101.

HUSTED, W. M. "A Rock Alignment in the Colorado Front Range," *Plains Anthropologist*, 8 (1963), p. 221.

KEHOE, T. F. "Stone 'Medicine Wheels' in Southern Alberta and the Adjacent Portion of Montana: Were They Designed as Grave Markers?" *Journal of the Washington Academy of Science*, 44 (1954), pp. 133–37.

KRUPP, E. C. "Cahokia: Corn, Commerce, and the Cosmos," *The Griffith Observer* (Los Angeles), vol. 41, no. 5 (May 1977), pp. 10–20.

————. "Sun and Stones on Medicine Mountain," *The Griffith Observer* (Los Angeles), vol. 38, no. 11 (November 1974), pp. 9–20.

MACGOWAN, KENNETH, and J. A. HESTER, JR. *Early Man in the New World.* Garden City, N.Y.: Doubleday & Company, 1962.

MARSHACK, ALEXANDER. *The Roots of Civilization.* New York: McGraw-Hill Book Company, 1972.

MILLER, WILLIAM C. "Two Prehistoric Drawings of Possible Astronomical Significance," *Astronomical Society of the Pacific Leaflet*, no. 314, July 1955.

PARSONS, ELSIE C. *Pueblo Indian Religion.* 2 vols. Chicago: University of Chicago Press, 1939.

REYMAN, JONATHAN. "Mexican Influence on Southwestern Ceremonialism." Ph.D. thesis. Southern Illinois University, 1971.

————. "Two Possible Solstitial Alignments at Pueblo Bonito, Chaco Canyon, New Mexico." Unpublished report, 1975.

————. "Astronomy, Architecture, and Adaptation at Pueblo Bonito," *Science*, 193 (1976), pp. 957–62.

ROBINSON, L. J. "Astronomy in Anthropology," *Astronomical Society of the Pacific Leaflet*, no. 380, February 1961.

SILVERBERG, ROBERT. *Home of the Red Man*. New York: Washington Square Press, 1971.
———. *The Mound Builders*. New York: Ballantine Books, 1970.
WEDEL, WALDO R. "The Council Circles of Central Kansas: Were they Solstice Registers?" *American Antiquity*, 32 (1967), pp. 54–63.
———. *Prehistoric Man on the Great Plains*. Norman: University of Oklahoma Press, 1961.
WILLIAMSON, R. A., et al. "The Astronomical Record in Chaco Canyon, New Mexico," in A. F. Aveni, ed., *Archaeoastronomy in Pre-Columbian America*. Austin: University of Texas Press, 1975. Pp. 33–44.
WITTRY, WARREN L. "An American Woodhenge," *Explorer*, 12 (1970), 4, pp. 14–17.
WORMINGTON, H. M. *Ancient Man in North America*. Denver, Col.: Denver Museum of Natural History, Popular Series No. 4, 1957.
———. *Prehistoric Indians of the Southwest*. Denver, Col.: Denver Museum of Natural History, Popular Series No. 7, 1947.
———, and R. G. FORBIS. "An Introduction to the Archaeology of Alberta, Canada." *Proceedings, Denver Museum of Natural History*, no. 11, 1965.

CHAPTER 5

ANDERSON, A. J. O., and C. E. DIBBLE, trs. (from Aztec to English). *Florentine Codex: General History of the Things of New Spain*, bk. 7. Santa Fe, N.M.: University of Utah and School of American Research, 1953.
AVENI, ANTHONY F. "Possible Astronomical Orientations in Ancient Mesoamerica," in A. F. Aveni, ed., *Archaeoastronomy in Pre-Columbian America*. Austin: University of Texas Press, 1975. Pp. 165–90.
———. *Native American Astronomy*. Austin: University of Texas Press, 1977.
———, and SHARON L. GIBBS. "On the Orientation of Precolumbian Buildings in Central Mexico," *American Antiquity*, vol. 41, (1976), pp. 510–17.
AVENI, ANTHONY F., and ROBERT M. LINSLEY. "Mound J, Monte Albán: Possible Astronomical Orientation," *American Antiquity*, vol. 37 (1972), pp. 528–31.
AVENI, ANTHONY F., SHARON L. GIBBS, and HORST HARTUNG. "The Caracol of Chichén Itzá—An Astronomical Observatory?" *Science*, 188 (1975), pp. 977–85.
CASO, A. *The Aztecs: People of the Sun*. Norman: University of Oklahoma Press, 1958.
———. *Exploraciones en Oaxaca*. Instituto Panamericano de Geografía e Historia, Publication 34, 1938.
COE, MICHAEL. "Native Astronomy in Mesoamerica," in A. F. Aveni, ed., *Archaeoastronomy in Pre-Columbian America*. Austin: University of Texas Press, 1975. Pp. 3–32.
DOW, JAMES. "Astronomical Orientation at Teotihuacán: A Case Study in Astroarchaeology," *American Antiquity*, vol. 32 (1967), pp. 326–34.
FLANNERY, KENT. "The Cultural Evolution of Civilizations," *Annual Review of Ecological Systems*, vol. 3 (1972), p. 399.
HARTUNG, HORST. *Die Zeremonialzentren der Maya*. Graz, Austria: Akademische Druck- und Verlaganstalt, 1971.
KRUPP, E. C. "The Observatory of Kukulkan," *The Griffith Observer* (Los Angeles), vol. 41, no. 9 (September 1977), pp. 1–20.
LANDA, DIEGO DE. *Relación de las Cosas de Yucatán*. Intro. and notes by H. Pérez Martínez. Mexico City, 1938.
LEVI-STRAUSS, CLAUDE. *The Raw and the Cooked*. New York: Harper & Row, 1969.
MARCUS, JOYCE. "Territorial Organization of the Lowland Classic Maya," *Science*, vol. 180 (1973), pp. 911–16.
MAUDSLAY, ALFRED P. *Archaeology: Biología Centrali Americana*. London: R. H. Porter and Dalau and Company, 1889–1902.

————. "A Note on the Position and Extent of the Great Temple Enclosure of Tenochtitlán," *Proceedings of the Eighteenth International Congress of Americanists*, London (1912), pp. 173–75.

MILLON, R. *Urbanization at Teotihuacán*. 2 vols. Austin: University of Texas Press, 1974.

MOTOLINÍA, TORIBIO. *Historia de los Indios de la Nueva España*. Mexico City: Chavez Hayhoe, 1941.

NUTTALL, ZELIA. "Nouvelles Lumières sur les Civilisations Américaines et le Système du Calandrier" (1906), *Atti del XXII Congresso Internazionale degli Americanisti Roma* (1926), pp. 119–48.

RICKETSON, OLIVER. *Carnegie Institute Washington Yearbook*, 28 (1925), p. 265.

RUPPERT, KARL. *The Caracol of Chichén Itzá, Yucatán, Mexico*. Washington, D.C.: Carnegie Institute Publication, no. 454, 1935.

SAHAGÚN, BERNARDINO DE. *Historia General de las Cosas de Nueva España* (1613), vols. 8–11. Ed. and tr. A. M. Garibay K. Mexico City: Biblioteca Porrúa, 1956.

STEPHENS, JOHN L. *Incidents of Travel in Central America, Chiapas and Yucatán* (1841). 2 vols. New York: Dover Publications, 1969.

————. *Incidents of Travel in Yucatán* (1843). 2 vols. New York: Dover Publications, 1963.

TEZOZÓMOC, ALVARADO H. *Crónica Mexicana* (1598). Mexico City: Editorial Leyenda, 1944.

THOMPSON, J. ERIC S. "A Commentary on the Dresden Codex, a Maya Hieroglyphic Book." *Memoirs of the American Philosophical Society* (1972), p. 93.

————. "A Survey of the Northern Maya Area," *American Antiquity*, vol. 1 (1945), p. 10.

————. *Maya Hieroglyphic Writing*. Norman: University of Oklahoma Press, 1971.

TICHY, FRANZ, "Deutung von Orts- und Flurnetzen im Hochland von Mexiko als Kulturreligiöse Reliktformen Altindianischer Besiedlung," *Erdkunde*, vol. 28 (1974), no. 3, pp. 194–207.

TORQUEMADA, JUAN DE. *Los Vientie i un Libros Rituales i Monarchia Indiana* (1615). 3 vols. Facsimile of 2d ed. (1723). Mexico City: Editorial Leyenda, 1943–44.

CHAPTER 6

ALDRED, CYRIL. *The Egyptians*. New York: Frederick A. Praeger, 1961.

BADAWY, ALEXANDER. "The Stellar Destiny of Pharoah and the So-Called Air-Shafts of Cheops' Pyramid," *Mitteilungen des Instituts für Orientforschung*, Band X (1964), pp. 189–206.

BAROCAS, C. *Monuments of Civilization Egypt*. New York: Grosset & Dunlap, 1972.

BLACKER, CARMEN, and MICHAEL LOEWE. *Ancient Cosmologies*. London: George Allen & Unwin, 1975.

BLACKMAN, A. M. *Luxor and Its Temples*. London: A. & C. Black, 1923.

BUDGE, E. A. WALLIS. *The Gods of the Egyptians*. 2 vols. New York: Dover Publications, 1969.

CASSON, LIONEL. *Ancient Egypt*. New York: Time-Life Books, 1965.

CHAMPOLLION, JACQUES. *The World of the Egyptians*. Tr. by J. Rosenthal. Geneva: Minerva S.A., 1971.

COTSWORTH, MOSES B. *The Fixed "Yearal" Proposed to Replace Changing Almanaks and Calendars*. New Westminster, B.C., Can.: International Almanak Reform League, 1914.

DE SANTILLANA, GIORGIO, and HERTHA VON DECHEND. *Hamlet's Mill*. Boston: Gambit, 1969.

EDWARDS, I. E. S. *The Pyramids of Egypt*. Harmondsworth, Eng.: Penguin Books, 1961.

ERMAN, ADOLF. *The Ancient Egyptians*. New York: Harper Torchbooks, 1966.

FAGAN, BRIAN M. *The Rape of the Nile*. New York: Charles Scribner's Sons, 1975.

FAGAN, CYRIL. *Zodiacs Old and New*. London: Anscombe & Company, 1951.

————. *Astrological Origins*. St. Paul, Minn.: Llewellyn Publications, 1971.

FAKHRY, AHMED. *The Pyramids*. Chicago: University of Chicago Press, 1961.

FRANKFORT, HENRI. *Ancient Egyptian Religion*. New York: Harper Torchbooks, 1961.

GLEADOW, RUPERT. *The Origin of the Zodiac*. New York: Castle Books, 1968.

GRINSELL, LESLIE V. *Barrow, Pyramid and Tomb*. London: Thames and Hudson, 1975.

HAWKES, JACQUETTA. *Atlas of Ancient Archaeology*. New York: McGraw-Hill Book Company, 1974.

HAWKINS, GERALD S. *Beyond Stonehenge*. New York: Harper & Row, 1973.

————. "Astronomical Alignments in Britain, Egypt, and Peru," *Philosophical Transactions of the Royal Society of London*, vol. 276 ("The Place of Astronomy in the Ancient World," ed. F. R. Hodson), 1974, pp. 157–67.

————. "Astroarchaeology: The Unwritten Evidence," in A. F. Aveni, ed., *Archaeoastronomy in Pre-Columbian America*. Austin: University of Texas Press, 1975.

————. "Stargazers of the Ancient World," 1976 *Yearbook of Science and the Future*. Chicago: Encyclopaedia Britannica, 1975. Pp. 124–37.

IONS, VERONICA. *Egyptian Mythology*. London: Paul Hamlyn, 1968.

LAUER, JEAN PHILIPPE. *Le Mystère des Pyramides*. Paris: Presses de la Cité, 1974.

LOCKYER, J. NORMAN. *The Dawn of Astronomy* (1894). Reprint edition, Cambridge, Mass.: M.I.T. Press, 1973.

MACQUITTY, WILLIAM. *Island of Isis*. New York: Charles Scribner's Sons, 1976.

MENDELSSOHN, KURT. *The Riddle of the Pyramids*. New York: Praeger Publishers, 1974.

MERCER, SAMUEL A. B. *Earliest Intellectual Man's Idea of the Cosmos*. London: Luzac & Company, 1957.

NEUGEBAUER, OTTO. *The Exact Sciences in Antiquity*. New York: Harper Torchbooks, 1962.

————. *A History of Ancient Mathematical Astronomy*, Part 2. New York: Springer-Verlag, 1975.

PANNEKOEK, A. *A History of Astronomy*. New York: Interscience Publishers, 1961.

PARKER, R. A. "Ancient Egyptian Astronomy," *Philosophical Transactions of the Royal Society of London*, vol. 276 ("The Place of Astronomy in the Ancient World," ed. F. R. Hodson), 1974, pp. 51–65.

PLUNKET, EMMELINE H. *Ancient Calendars and Constellations*. London: John Murray, 1903.

PROCTOR, RICHARD A. *Myths and Marvels of Astronomy*. London: Longmans, Green, and Company, 1889.

SARTON, GEORGE. *A History of Science*, vol. I. New York: W. W. Norton & Company, 1970. Reprint of 1952 edition.

SMYTH, C. PIAZZI. *Our Inheritance in the Great Pyramid*. London: Charles Burnet and Company, 1890.

TOMPKINS, PETER. *Secrets of the Great Pyramid*. New York: Harper & Row, 1971.

TRIMBLE, VIRGINIA. "Astronomical Investigation Concerning the So-Called Air-Shafts of Cheops' Pyramid," *Mitteilungen des Instituts für Orientforschung*, Band X (1964), pp. 183–87.

WAKE, C. STANILAND. *The Origin and Significance of the Great Pyramid* (1882). Reprint edition, Minneapolis: Wizards Bookshelf, 1975.

WOOD, MARGOT. "From Nile to Nineveh: Calendars, Pyramids, and Star Fortunes." Talk given at Los Angeles February 25, 1975, and San Diego February 26, 1975, for University of California Extension course "In Search of Ancient Astronomies."

CHAPTER 7

When Worlds Collide: The Velikovsky Catastrophe

ASIMOV, ISAAC. "CP," *Analog*, vol. XCIV, no. 2 (October 1974), pp. 38–50.
————. *The Stars in Their Courses*. New York: Ace Books, 1972. Pp. 45–56, "Worlds in Confusion."
BOVA, BEN. "The Whole Truth," *Analog*, vol. XCIV, no. 2 (October 1974), pp. 5–11.
BRACEWELL, RONALD N. *The Galactic Club: Intelligent Life in Outer Space*. San Francisco: W. H. Freeman and Company, 1975. Pp. 17–25, "Velikovskian Vermin."
COHEN, DANIEL. *Myths of the Space Age*. New York: Dodd, Mead & Company, 1967. Pp. 172–94, "Immanuel Velikovsky—the Man Who Challenged the World."
Editors of *Pensée*. *Velikovsky Reconsidered*. Garden City, N.Y.: Doubleday & Company, 1976.
GARDNER, MARTIN. *Fads and Fallacies in the Name of Science*. New York: Dover Publications, 1957. Pp. 28–41, "Monsters of Doom."
GRAZIE, ALFRED DE. *The Velikovsky Affair*. New York: University Books, 1966.
JUENEMAN, FREDERIC B. "The Search for Truth," *Analog*, vol. XCIV, no. 2 (October 1974), pp. 25–37.
MACKIE, E. W. "Megalithic Astronomy and Catastrophism," *Pensée*, vol. 4 (1974), pp. 5–20.
Pensée, vol. 4 (1974), no. 2, report on AAAS Symposium and related commentaries.
RUSSELL, JOHN A. Book review of *Worlds in Collision*, *Navigation*, vol. 2, no. 8 (December 1950), pp. 294–95.
"Scientists in Collision," *Newsweek*, February 25, 1974, pp. 58–59.
SLADEK, JOHN. *The New Apocrypha*. New York: Stein and Day Publishers, 1974. Pp. 19–30, "Before the Invasion."
VELIKOVSKY, IMMANUEL. *Ages in Chaos*, vol. 1. Garden City, N.Y.: Doubleday & Company, 1952.
————. *Earth in Upheaval*. Garden City, N.Y.: Doubleday & Company, 1955.
————. *Peoples of the Sea*. Garden City, N.Y.: Doubleday & Company, 1977.
————. *Worlds in Collision*. New York: The Macmillan Company, 1950.
"Velikovsky and the AAAS: Worlds in Collision," *Science News*, March 2, 1974, p. 132.
WARSHOFSKY, FRED. "When the Sky Rained Fire: The Velikovsky Phenomenon," *Reader's Digest*, December 1975, pp. 219–40.

Ley or Nay? and *The Glastonbury Zodiac*

BORD, JANET, and COLIN BORD. *Mysterious Britain*. Garden City, N.Y.: Doubleday & Company, 1973.
————. *The Secret Country*. London: Paul Elek, 1976.
CAINE, MARY. "The Glastonbury Zodiac," *The News*, no. 4 (1974), pp. 8–14. (Reprinted and revised from its original appearance in *Gandalf's Garden*, no. 4, 1969). See below, under *Miscellaneous*.
EITEL, E. J. *Feng Shui* (1873). Cambridge, Eng.: The Land of Cokaygne, 1973.
FORREST, ROBERT. "Leys, UFOs, and Chance," *The News*, no. 13 (December 1975), pp. 12–13. See below, under *Miscellaneous*.
HITCHING, FRANCIS. *Earth Magic*. New York: William Morrow and Company, 1977.

HOLROYD, STUART. *Magic, Words, and Numbers*. Garden City, N.Y.: Doubleday & Company, 1976.

KING, FRANCIS. *Wisdom from Afar*. Garden City, N.Y.: Doubleday & Company, 1976.

KRUPP, E. C. "Ley or Nay," *Stonehenge Viewpoint*, vol. 8 (1976), no. 3, p. 3.

LOCKYER, J. NORMAN. *Stonehenge and Other British Stone Monuments Astronomically Considered*. London: Macmillan & Company, 1909.

MALTWOOD, K. E. *A Guide to Glastonbury's Temple of the Stars*. London: James Clarke & Co., 1964. Reprint of 1929 edition.

MICHELL, JOHN. *City of Revelation*. London: Garnstone Press, 1972.

————. *The Flying Saucer Vision*. London: Abacus, 1974.

————. *The Old Stones of Land's End*. London: Garnstone Press, 1974.

————. *The View over Atlantis*. London: Garnstone Press, 1975.

PENNICK, NIGEL. *Geomancy*. Cambridge, Eng.: Cokaygne Publishing, 1973.

————. "The Geomancy of Glastonbury Abbey," *Megalithic Vision Antiquarian Papers*, no. 11. Cambridge, Eng.: Fenris Wolf, 1976.

————. "Leys and Zodiacs," *Megalithic Visions Antiquarian Papers*, no. 5. Cambridge, Eng.: Fenris Wolf, 1975.

REISER, OLIVER L. *This Holyest Erthe*. London: Perennial Books, 1974.

ROBERTS, ANTHONY. *Atlantean Traditions in Ancient Britain*. Llanfynydd, Carmarthen, Wales: Unicorn Bookshop, 1974.

————. *Glastonbury: Ancient Avalon, New Jerusalem*. London: Zodiac House Publications, 1976.

SCREETON, PAUL. *Quicksilver Heritage*. Wellingborough, Northamptonshire: Thorsons Publishers, 1974.

TRENCH, BRINSLEY LE POER. *Men Among Mankind*. London: Neville Spearman, 1962.

UNDERWOOD, GUY. *The Pattern of the Past*. London: Abacus, 1974.

WATKINS, ALFRED. *The Old Straight Track*. London: Garnstone Press, 1974. Reprint of 1925 edition.

WILLIAMS, MARY, ed. *Britain, a Study in Patterns*. London: Research into Lost Knowledge Organization, 1971.

————. *Glastonbury, a Study in Patterns*, London: Research into Lost Knowledge Organization, 1969.

Miscellaneous

The following periodicals regularly carry unorthodox interpretations of antiquities and are available by subscription:

The Ley Hunter (Paul Devereux, P. O. Box 152, London, N10 1EP, England.)

Fortean Times (formerly *The News*) (Robert J. M. Rickard, Post Office Box 152, London N10 1EP, England.)

Stonehenge Viewpoint (Donald L. Cyr, Post Office Box 30887, Santa Barbara, California 93105, U.S.A.

A Serious Mystery

KRUPP, E. C. "On Not Taking It Seriously," *The Griffith Observer* (Los Angeles), vol. 40, no. 9 (September 1976), pp. 16–17.

LOCKYER, J. NORMAN. *The Dawn of Astronomy* (1894). Reprint edition, Cambridge, Mass.: M.I.T. Press, 1973.

SEVER, TOM. "The Obsession with the Star Sirius," *The Griffith Observer* (Los Angeles), vol. 40, no. 9 (September 1976), pp. 8–15.

TEMPLE, ROBERT K. G. *The Sirius Mystery*. New York: St. Martin's Press, 1976.

Assorted Ancient Astronauts

BELLAMY, HANS SCHINDLER, and PETER ALLEN. *The Great Idol of Tiahuanaco.* London: Faber & Faber, 1959.

BERGIER, JACQUES. *Extraterrestrial Visitations from Prehistoric Times to the Present.* New York: Signet Books, 1974.

———, and the Editors of *Info. Extraterrestrial Intervention: the Evidence.* New York: Signet Books, 1975.

BERLITZ, CHARLES. *Mysteries from Forgotten Worlds.* New York: Dell Publishing Co., 1973.

BLUMRICH, JOSEF F. *The Spaceships of Ezekiel.* New York: Bantam Books, 1974.

CHARROUX, ROBERT. *Forgotten Worlds.* New York: Popular Library, 1973.

———. *The Gods Unknown.* New York: Berkley Medallion Books, 1974.

———. *Legacy of the Gods.* New York: Berkley Medallion Books, 1974.

———. *Masters of the World.* New York: Berkley Medallion Books, 1974.

———. *The Mysterious Past.* New York: Berkley Medallion Books, 1975.

———. *One Hundred Thousand Years of Man's Unknown History.* New York: Berkley Medallion Books, 1971.

COHEN, DANIEL. *The Ancient Visitors.* Garden City, N.Y.: Doubleday & Company, 1976.

COLLYNS, ROBIN. *Did Spacemen Colonise the Earth?* Frogmore, St. Albans, Herts., England: Mayflower Books, Ltd., 1974.

———. *Laser Beams from Star Cities.* London: Sphere Books, 1977.

DÄNIKEN, ERICH VON. *Chariots of the Gods?* Tr. by Michael Heron. New York: G. P. Putnam's Sons, 1970.

———. *Gods from Outer Space.* Tr. by Michael Heron. New York: G. P. Putnam's Sons, 1971.

———. *The Gold of the Gods.* Tr. by Michael Heron. New York: G. P. Putnam's Sons, 1973.

———. *In Search of Ancient Gods.* Tr. by Michael Heron. New York: G. P. Putnam's Sons, 1974.

———. *Miracles of the Gods.* Tr. by Michael Heron. London: Souvenir Press, 1976.

DILLON, JOHN, et al. "Lost Worlds and Golden Ages." Third Annual Division of Interdisciplinary and General Studies Interdisciplinary Symposium, University of California, Berkeley, May 30, 1974. Tapes available through University of California Extension Media Center, Berkeley, California 94720.

DOWNING, BARRY H. *The Bible and Flying Saucers.* New York: Avon Books, 1970.

DRAKE, W. RAYMOND. *Gods and Spacemen in the Ancient East.* New York: Signet Books, 1968.

———. *Gods and Spacemen in the Ancient Past.* New York: Signet Books, 1974.

———. *Gods and Spacemen in the Ancient West.* New York: Signet Books, 1974.

———. *Gods and Spacemen in Greece and Rome.* New York: Signet Books, 1977.

———. *Gods and Spacemen Throughout History.* Chicago: Henry Regnery Company, 1975.

———. *Gods or Spacemen?* New York: Signet Books, 1976.

FERRIS, TIMOTHY. "Playboy Interview: Erich von Däniken," *Playboy,* vol. 21, no. 8, August 1974, pp. 51ff.

FLINDT, MAX H., and OTTO O. BINDER. *Mankind—Child of the Stars.* Greenwich, Conn.: Fawcett Publications, 1974.

GINSBURGH, IRWIN. *First, Man. Then, Adam!* New York: Simon and Schuster, 1977.

GUPTON, JAMES A. "Ancient Astronauts Left Us a Legacy," *Beyond Reality,* no. 21 (July–August 1976), pp. 31–32.

HUTIN, SERGE. *Alien Races and Fantastic Civilizations.* New York: Berkley Medallion Books, 1975.

KOLOSIMO, PETER. *Not of This World.* New York: Bantam Books, 1973.

———. *Spaceships in Prehistory.* Secaucus, N.J.: University Books, 1976.

————. *Timeless Earth*. New York: Bantam Books, 1975.

KRUPP, E. C. "The von Däniken Phenomenon," *The Griffith Observer* (Los Angeles), vol. 38, no. 4 (April 1974), pp. 2–14. (Reprinted vol. 41, no. 7 [July 1977])

LANDSBURG, ALAN, and SALLY LANDSBURG. *In Search of Ancient Mysteries*. New York: Bantam Books, 1974.

————. *The Outer Space Connection*. New York: Bantam Books, 1975.

LETHBRIDGE, T. C. *The Legend of the Sons of God*. London: Routledge & Kegan Paul, 1972.

LEWIN, L. M. *Footprints on the Sands of Time*. New York: Signet Books, 1975.

LINGEMAN, RICHARD R. "Erich von Däniken's Genesis," *New York Times Book Review*, March 30, 1974, p. 6.

MACHLIN, MILT. "Ancient Space Visitors Revealed by Erich von Däniken, Author of *Chariots of the Gods*," *Argosy*, June 1974, pp. 54ff.

MCLEAN, R. GORDEN. *Mementos from the Universe*. New York: Carlton Press, 1976.

MCNARY, DAVE. "Were There Ancient Astronauts (or Is This a Put-on)?" *U.C.L.A. Daily Bruin*, April 10, 1974.

MELTZER, EDMUND S. "Swing Lower, Sweet Chariots of the Gods!" *Fate* (Highland Park, Ill.), vol. 29, no. 7 (July 1976), pp. 34–42.

MOONEY, RICHARD E. *Colony Earth*. Greenwich, Conn.: Fawcett Publications, 1975.

————. *Gods of Air and Darkness*. Greenwich, Conn.: Fawcett Publications, 1975.

NORMAN, ERIC. *Gods, Demons, and Space Chariots*. New York: Lancer Books, 1970.

————. *Gods and Devils from Outer Space*. New York: Lancer Books, 1973.

OSTRIKER, ALICIA. "What If We're Still Scared, Bored and Broke?" *Esquire*, December 1973, pp. 238ff.

PAUWELS, LOUIS, and JACQUES BERGIER. *The Eternal Man*. New York: Avon Books, 1973.

————. *The Morning of the Magicians*. New York: Avon Books, 1973.

SEGRAVES, KELLY L. *Sons of God Return*. New York: Pyramid Books, 1975.

SENDY, JEAN. *The Coming of the Gods*. New York: Berkley Medallion Books, 1973.

————. *The Moon: Outpost of the Gods*. New York: Berkley Medallion Books, 1975.

————. *Those Gods Who Made Heaven and Earth*. New York: Berkley Medallion Books, 1972.

SITCHIN, ZECHARIA. *The 12th Planet*. New York: Stein & Day, 1976.

STEIGER, BRAD. *Mysteries of Time and Space*. New York: Dell Publishing Company, 1976.

STEINHAUSER, GERHARD R. *Jesus Christ: Heir to the Astronauts*. New York: Pocket Books, 1976.

STORY, RONALD. *The Space-Gods Revealed*. New York: Harper & Row, 1976.

THIERING, BARRY, and EDGAR CASTLE. *Some Trust in Chariots*. New York: Popular Library, 1972.

THOMAS, ROY, and MARIE SEVERIN. "Hot Rods of the Gods," *Crazy*, no. 3, March reprint).

THOMAS, ROY, and MARIE SEVERIN. "Hot Rods of the Gods," *Crazy*, no. 3, March 1974, pp. 29–34.

TOMAS, ANDREW. *Beyond the Time Barrier*. New York: Berkley Medallion Books, (1974, 1976 reprint).

————. *The Home of the Gods*. New York: Berkley Medallion Books, (1972, 1974 reprint).

————. *On the Shores of Endless Worlds*. Bantam Books, Inc., (1974, 1976 reprint).

————. *We Are Not the First*. New York: Bantam Books, Inc., (1971, 1973 reprint).

TRENCH, BRINSLEY LE POER. *The Sky People*. New York: Award Books, (1960, 1975 reprint).

————. *Temple of the Stars (Men among Mankind)*. New York: Ballantine Books, (1962, 1974 reprint).

UMLAND, ERIC, and CRAIG UMLAND. *Mystery of the Ancients (Early Spacemen and the Mayas)*. New York: Signet Books, (1974, 1975 reprint).

VAN DER VEER, M. H. J., and P. MOERMAN. *Hidden Worlds*. New York: Bantam Books, Inc., 1973.

WHITE, PETER. *The Past Is Human*. New York: Taplinger Publishing Company, 1976.

WILLIAMSON, GEORGE HUNT. *Road in the Sky*. London: Futura Publications, 1975.

————. *Secret Places of the Lion*. Futura Publications, 1974.

WILSON, CLIFFORD. *Crash Go the Chariots*. New York: Lancer Books, 1972.

————. *The Chariots Still Crash*. New York: Signet Books, 1975.

Miscellaneous

Comic-book versions of the ancient astronaut theme included:

Man-Gods from Beyond the Stars (Marvel Presents ⚡1, 1975)

Tragg and the Sky Gods (Gold Key, bimonthly)

The Eternals (Marvel Comics, monthly)

Ancient Astronauts, published bimonthly since January 1976 (Winter 1975), Countrywide Publications, 257 Park Avenue South, New York, New York 10010. An entire magazine devoted to the ancient astronaut theme.

The Griffith Observer, published monthly by Griffith Observatory, 2800 East Observatory Road, Los Angeles, California 90027. Regularly features articles of archaeoastronomical interest as well as many articles on modern astronomy and related sciences. Available by subscription from Griffith Observatory ($5.00, one year; $9.00, two years; $13.00, three years).

The Sourcebook Project, published by William R. Corliss, Glen Arm, Maryland 21057. Includes several volumes of interest to readers of this book. *Strange Artifacts*, vols. 1 and 2, include a number of reprints of rare or obscure papers on ancient monuments, and in many instances archaeoastronomical interpretations are discussed. Publication began 1974.

Index